I0056306

CHINESE CYMBIDIUM ORCHID

A Gentleman of Noble Virtue

CHINESE CYMBIDIUM ORCHID

A Gentleman of Noble Virtue

Hew Choy Sin

Wong Yik Suan

World Scientific

NEW JERSEY · LONDON · SINGAPORE · BEIJING · SHANGHAI · HONG KONG · TAIPEI · CHENNAI · TOKYO

Published by

World Scientific Publishing Co. Pte. Ltd.

5 Toh Tuck Link, Singapore 596224

USA office: 27 Warren Street, Suite 401-402, Hackensack, NJ 07601

UK office: 57 Shelton Street, Covent Garden, London WC2H 9HE

Library of Congress Cataloging-in-Publication Data
Names: Hew, Choy Sin, author. | Wong, Yik Suan, author.
Title: Chinese cymbidium orchid : a gentleman of noble virtue / Choy Sin Hew ; Wong Yik Suan
Description: First edition | Hackensack, NJ : World Scientific, [2024] |
    Includes bibliographical references and index.
Identifiers: LCCN 2022045048 | ISBN 9789811263361 (hardcover) |
    ISBN 9789811263378 (ebook) | ISBN 9789811263385 (ebook other)
Subjects: LCSH: Orchid culture--China--History. | Cymbidium--Physiology--China |
    Cymbidium--History--China.
Classification: LCC SB409.8.C95 H49 2024 | DDC 635.9/34720951--dc23/eng/20220928
LC record available at https://lccn.loc.gov/2022045048

British Library Cataloguing-in-Publication Data
A catalogue record for this book is available from the British Library.

Copyright © 2024 by World Scientific Publishing Co. Pte. Ltd.

All rights reserved. This book, or parts thereof, may not be reproduced in any form or by any means, electronic or mechanical, including photocopying, recording or any information storage and retrieval system now known or to be invented, without written permission from the publisher.

For photocopying of material in this volume, please pay a copying fee through the Copyright Clearance Center, Inc., 222 Rosewood Drive, Danvers, MA 01923, USA. In this case permission to photocopy is not required from the publisher.

For any available supplementary material, please visit
https://www.worldscientific.com/worldscibooks/10.1142/13056#t=suppl

Typeset by Stallion Press
Email: enquiries@stallionpress.com

This book serves as a brief introduction to the history of the Chinese Cymbidium orchid. It describes appreciation and praise of the Chinese Cymbidium orchid by ancient scholars and artists both in the Eastern and Western civilization. The authors share reprints of written articles and paintings and calligraphies of the Chinese Cymbidium orchids, traced back as early as the Spring and Autumn period (770-476 BC) in China. An interesting observation is that the Chinese painters draw the entire Chinese Cymbidum orchid plant that includes its leaves, while the Western artists seem to be attracted more by the size, form and colour of the orchid flower itself. The morphology and biology of five common Chinese Cymbidium orchids are described with beautiful photographic images. A brief introduction on how the orchid is commercially propagated and cultivated in the past and now. Different symptoms of its common diseases and pest management are presented. The Chinese Cymbidium orchid symbolizes elegance, nobility, resilience and purity. Therefore, it is regarded as the botanical treasure of China.

I highly recommend this book to readers who wish to learn more about Chinese Cymbidium orchid. The knowledge of Chinese Cymbidium orchid will enrich our appreciation for its beauty and the virtues it symbolizes into our real life.

Professor Wong Sek Man
Department of Biological Sciences
National University of Singapore

For our grandchildren Chien Ying, Kean Yee,
Shin Han and Shih Yee

Preface

The cultivation of Chinese cymbidiums has a long history in China dating back more than a thousand years. Chinese cymbidiums are commonly known as "*Lan*" (兰) or "*Guo Lan*" (国兰) in China. It refers to a group of terrestrial cymbidiums belonging to the subgenera *Jensoa* of Genus *Cymbidium*. The best-known Chinese cymbidiums include *C. goeringii* (春兰), *C. sinense* (墨兰), *C. ensifolium* (建兰), *C. faberi* (蕙兰), and *C. kanran* (寒兰).

Lan has a special appeal and meaning to the Chinese as it symbolises integrity, modesty, and nobility. Confucius was full of praise for *Lan* as the "*Gentleman of Noble Virtue*" (君子之风) and "*King of Fragrance*" (王者之香). The love and appreciation for *Lan*, also known as "*Lan Culture*", has become an integral part of Chinese culture.

We first came across Chinese cymbidiums when visiting the Guangzhou Flower Research Center in 1986. We were fascinated by the charm and beauty of these orchids especially its exotic flowers and sweet fragrance. This has motivated us to read more of the history and culture of Chinese cymbidiums, and to visit research laboratories and orchid nurseries in and around Guangzhou, Kunming, and Nanjing. In 1987, collaborative research on the physiology of Chinese cymbidiums was initiated between the National University of Singapore and South China Normal University in Guangzhou, China.

The COVID-19 pandemic is creating havoc and disruption in our daily lives. To make good use of our time during the lockdown, we decided to start writing a book on Chinese cymbidiums, an idea we have had for some years. The aim of writing this book is to share with you the joy we have had in knowing these beautiful and graceful orchids.

This book is written as an introduction to the history, literature, art, traditional customs, biology, and cultivation of Chinese cymbidiums, and their association with Chinese culture. We have conducted extensive desk research to locate all relevant information available in books (including classical Chinese books), periodicals, and the Internet in all related areas, but omission is inevitable. We hope the readers will find this introductory book interesting and be motivated to read more about Chinese cymbidiums and the *Lan Culture* in China.

We are grateful for all the assistance we received during the writing of this book. We would like to express our heartfelt gratitude to both Prof. Joseph Arditti, Professor of Biology Emeritus, University of California, Irvine, USA, and Prof. Wong Sek Man, Dept. of Biological Sciences, National University of Singapore, for their valuable comments and suggestions. Our very special thanks also go to Mr. Lim See Young, Singapore's renowned calligrapher, for the elegant piece of calligraphy on *Lan*; Prof. Ye Qingsheng of South China Normal University, China; Ms Lee Foong Ying of Singapore; and many other overseas friends for their help in securing and/or taking some of the photos for the book. We would also like to thank Prof. K.K. Phua, Chairman of World Scientific Publishing, for his encouragement and support, and Ms. Joy Quek, for her advice on editorial matters and assistance.

Hew Choy Sin
Wong Yik Suan
May 2022

About the Authors

Hew Choy Sin, former Professor of Botany from the National University of Singapore (NUS), is a leading authority on the physiology of tropical orchids, an area that he had been involved in for more than 35 years. Prof. Hew has published 3 books and more than 150 scientific papers. In 1997, Prof. Hew was awarded the Singapore National Science Award (the highest honour for a research scientist in Singapore) for his outstanding contribution in placing Singapore at the forefront of global orchid research.

In addition, he was formerly the Deputy Director and Director of Institute of Natural Sciences of Nanyang University (1973–1980), as well as an Adjunct Professor and Deputy Director of the Institute of Advanced Studies of Nanyang Technological University in Singapore (2005–2011).

Prof. Hew had previously served as an advisor to Guangzhou Scientific Technology Exchange Centre (China) on horticultural production. He was a Visiting Professor of multiple academic institutions including Nanjing Agricultural University, Xiamen University, South China Normal University, Taiwan National Cheng Kung University, and a Visiting Scholar of University of South Australia and Department of Scientific and Industrial Research (DSIR) in New Zealand.

Wong Yik Suan was former Senior Librarian at the National University of Singapore before her retirement in 2010. She obtained her B.Sc. in Biological Sciences at Nanyang University in 1971. She started as a librarian at Nanyang University and later continued her career at the National University of Singapore. She obtained her Postgraduate Diploma in Librarianship (UK) in 1982 under a NUS Staff Development Scholarship.

Acknowledgements

We would like to express our heartfelt gratitude to the following friends for their support and contribution in providing photos for the book, and verification of the names of some Chinese *Cymbidium* varieties.

Prof. Ye Qingsheng (叶庆生), School of Life Sciences, South China Normal University, Guangzhou, China.

Ms. Lee Foong Ying (李凤英), Edu-programme Consultant, Orchidville Pte Ltd, and Operation Manager, Bugs & Buds, Singapore.

Mr. Wang Jiaxing (王家兴), Ning Bi Yuan Orchid Research Institute (凝碧苑兰花研究所), Songyang County, Zhejiang, China.

Ms. Gao Lixia (高丽霞), Zhongkai University of Agriculture and Engineering, Guangzhou, China.

Prof. Zhu Genfa (朱根发), Environmental Horticulture Research Institute, Guangdong Academy of Agricultural Sciences, Guangzhou, China.

Prof. Xu Yechun (徐晔春), Environmental Horticulture Research Institute, Guangdong Academy of Agricultural Sciences, Guangzhou, China.

Ms. Yang Fengxi (杨凤玺), Environmental Horticulture Research Institute, Guangdong Academy of Agricultural Sciences, China.

Assoc. Prof. Wang Zaihua (王再花), Environmental Horticulture Research Institute, Guangdong Academy of Agricultural Sciences, China.

Ms. Yang Fengxi (杨凤玺), Environmental Horticulture Research Institute, Guangdong Academy of Agricultural Sciences, China.

Ms. Lily Lai (赖丽丽), Dongfang Tenfei Hort. Biotech. Ltd., Hainan, China.

Ms. Wang Miao Miao (王苗苗), Beijing Botanical Garden, Beijing, China.

Mr. Guo Sheng (郭盛), Longyan Sanhe Agritechnology Ltd., China.

Ms. Huang Su Nan (黄素南), Ever-Where Biot. Co., Taichung, Taiwan.

Mr. Huang Ming-zhong (黄明忠), Tropical Crops Genetic Resources Research Institute, Chinese Academy of Tropical Agricultural Sciences, Hainan, China.

Acknowledgements
(Photographs)

Gao Lixia (Figs. 4.10, 4.11, 5.2)

Guo Sheng (Figs. 4.10, 4.17)

Huang Su Nan (Figs. 3.21, 3.22, 3.26, 3.27, 3.28, 4.9, 4.10)

Lily Lai (Figs. 4.15, 4.17)

Lee Foong Ying (Figs. 4.10, 4.16(b)–4.16(f))

Lim See Young (Fig. 1.1)

Wang Jiaxing (Figs. 2.28, 3.41, 3.42, 3.43, 3.46, 3.49, 3.50)

Wang Miao Miao (Figs. 3.4, 3.5, 3.12, 3.19, 3.24, 3.28, 3.35, 3.45, 4.7, 4.10, 4.11)

Xu Yechun/Wang Zaihua (Figs. 3.2, 3.7, 3.8, 3.20, 3.23, 3.25, 3.33, 3.37, 3.38, 3.39, 3.44, 3.47, 3.48, 4.10, 4.12, 4.13, 4.18, 5.4)

Yang Fengxi (Figs. 3.15, 3.17, 4.10, 5.2, 6.1)

Ye Qingsheng (Figs. 3.29, 3.31, 3.32, 3.34, 3.40, 4.2, 4.3, 4.4, 4.14)

Contents

Chapter 1

History of Chinese *Cymbidium*

1.1. Introduction

The Orchid family (Orchidaceae) is one of the three largest families in the plant kingdom; the other two being the Asteraceae (Daisy family) and Poaceae (Grass family). The Orchid family has over 749 genera and 26,000 species. However, there are more than 100,000 horticultural hybrids, some of which no longer exist. Nevertheless, additional hybrids are being produced and registered with the Royal Horticultural Society in the U.K. almost daily. Orchids are monocotyledonous plants despite having only minimal or no cotyledons in their seeds and are regarded as one of the most highly evolved plant families.[2,3,17] Of the many orchid species, approximately 25% are terrestrial, 70% epiphytes, and the other 5% can grow on a variety of substrates. Orchids can be grouped into four main categories based on their natural habitats[2]:

- Terrestrial orchids (e.g., *Arundina; Cymbidium*)
- Epiphytic orchids (e.g., *Dendrobium; Phalaenopsis*)
- Lithophytic orchids (e.g., *Eria*)
- Saprophytic orchids (e.g., *Gastrodia*)

It is interesting to know the scientific term "*orchid*" is derived from the Greek word for "*orchis*" meaning "*testis*". *Orchis* refers specifically to the two tubers of *Orchis mascula* that resemble the mammalian male sex organ. It is generally agreed that the word "*orchis*" was first used by the Greek philosopher Theophrastus (370–285 BC) in his study of plants. The term "*orchid*", later retained by Carl Linnaeus, the founder of modern plant taxonomy in his "*Species Plantarum*", currently refers to the entire orchid family.[2,3]

The love and cultivation of orchids have a long history in both Western and Eastern cultures, but the appreciation of the beauty and mystery of orchids is very subjective. Traditionally, the Chinese appreciate orchid flowers rather differently from the Westerners due to disparities in aesthetic concepts and appreciation standards.

In the west, there is a strong interest and tradition in trying to understand the biodiversity, structural adaptation, survival, and evolution of orchids. Much of this is attributed to the extraordinarily diverse forms and structures of orchids. Some orchids resemble bees, moths, monkeys, dancing dolls, butterflies, slippers and many others. The structure of each orchid flower is unique and has undergone significant structural modification to ensure pollination and safeguard survival. The intriguing pollination phenomena of the orchid flower, for example, has attracted the attention of numerous biologists, including Charles Darwin and those of the like.[2,3]

The appreciation of orchids in the western hemisphere focuses largely on the flower itself — sizes, types, forms, and colours — while the leaf is less appreciated. The great diversity and beauty of the flower structure in the vast numbers of orchid species and hybrids are a strong attraction. Generally, western orchid flowers like *Cattleya* and *Phalaenopsis* are larger in size, colourful, gorgeous, and lovely. In contrast, Chinese *Cymbidium* flowers are not eye-catching or brightly coloured. They are small, simple and elegant, and mostly white, green and yellow. Most are endowed with a strong yet delicate scent. These features conform well to the aesthetic standards of Asian people. Apart from the noble qualities that the *Cymbidium* flower symbolises, the Chinese pay special attention to the overall beauty of the orchid plant, including the flower, leaf, stem and scent. Remarkably, even the degree of scent, plant form and the type of flowerpot used are taken into account and serve as criteria for judging.[60,61]

1.2. The Origin and Meaning of "*Lan*"

Chinese *Cymbidium* (also known as *Oriental orchid*) has a special place in the hearts of the Chinese, and they are generally known as "*Lan*" (兰) or "*Guo Lan*" (国兰). There were written records about *Lan* dating back more than two thousand years in the history of Chinese civilisation. According to historical records, *lan* in China was mentioned as early as the Spring and Autumn Period (春秋时代; 770–476 BC).[65,68] In the minds of the Chinese, *lan* has an oriental sense of mystery. It symbolises elegance, nobility and purity. Additionally, it can grow and survive in a very harsh environment and yet maintain its majestic air of sophistication and beauty, thus signifying resilience and nobility. As such, it is regarded as the botanical treasure of China.

Lan is mentioned in several Chinese classics texts, including "*Shi-Jing*" (诗経) and "*Li-Ji*" (礼记). *Shi-Jing (The Classic of Poetry or The Book of Songs)* is the earliest collection of Chinese poetry dating from the 11th to 7th century BC. *Li-Ji (The Book of Rites)* was most likely published during the Western Han era (206 BC–25 AD). In these books, Chinese cymbidiums were referred to as "*ni*" (鷁), "*chien*" (蕳), and "*lan*" (兰). However, "*ni*" refers to ribbon grass (绶草; *Spiranthes sinensis*) but not the actual Chinese *Cymbidium*.[4,65] In many ancient writings, "*chien*" and "*lan*" were used as a loose reference to several fragrant plants used in religious ceremonies to

---

*The chronology of the Chinese dynasties is based on the book *Reading Chinese Painting* by S.M. Law (2016).[22]

ward off evil spirits instead of being an exclusive reference to orchids. This resulted in much confusion regarding the exact dates for the beginning of the cultivation of Chinese cymbidiums in China.[65]

It is believed that Confucius (551–479 BC), the great Chinese philosopher and educator of the Spring and Autumn Period, was the first person to associate *lan* with Chinese culture. As the story goes, Confucius was travelling in various parts of China, attempting to promote his ideas of ethics and virtue as an ideal form of government but failed despite much effort. One day, as he was returning from Wei (卫) to Lu (鲁) and wandering in a valley, he came across some fragrant plants growing in the wild and commented[60,64]:

芝兰生于幽谷，不以无人而不芳；
君子修道立德，不以穷困而变节。

When translated literally, it means "*zhi lan*" (芝兰) "*that grows in deep valleys never withholds its fragrance, even when it is not being appreciated. Similarly, men of noble character will not let poverty affect their determination to be guided by high principles and morality.*"

Confucius further commented:

夫兰当为王者香,
今乃独茂, 与众草为伍。

This translates to "*Lan, regarded as the King of Fragrance, is living among wild grasses, but still growing well and lushly.*"

Moreover, he likened *lan* to a "*Gentleman*" because it is elegant, virtuous, and refined, symbolising a man of nobility and virtue (君子 之凤).[47,64] (Fig. 1.1) That was the beginning of relating *lan* to the human character, as Confucius believed it should form the basis of his moral teachings.

Confucius also said[64]:

与善人居，如入芝兰之室,
久而不闻其香，即与之化矣。

君子之風

王者之香

林書香

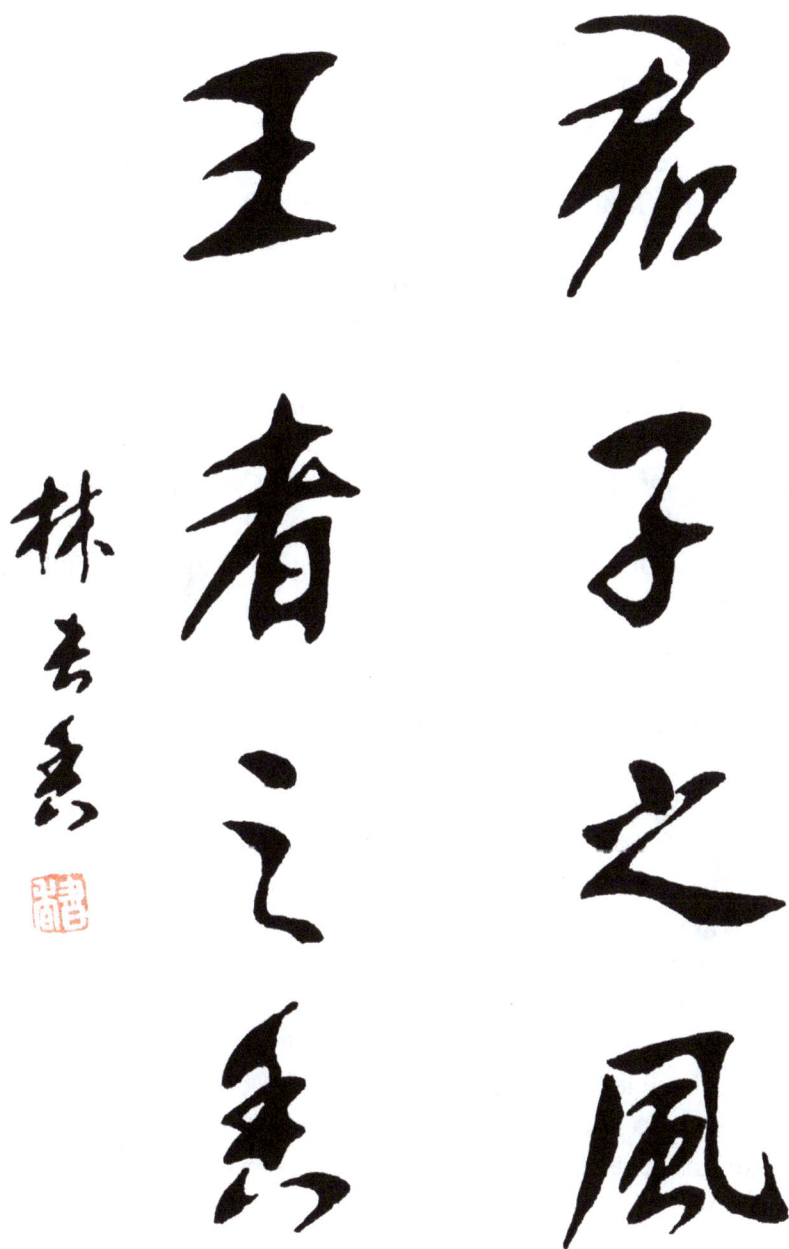

Fig. 1.1. "*A Gentleman of Noble Virtue and King of Fragrance*". Calligrapher: Ling See Young (林书香) (Singapore)

In it, he stressed the importance of getting acquainted with a person of high moral character and likened it to entering a room filled with the fragrance of *lan*. Over time, one does not detect or feel the fragrance anymore because one has become accustomed to it. In other words, by associating with people of noble character, one slowly learns from them and will be positively influenced by their personality.

Some scholars maintained that the *zhi lan* mentioned by Confucius was a true Chinese *Cymbidium* because the habitat as described by him was similar to that of wild Chinese cymbidiums today. This view, however, was not shared by most Chinese orchidologists. They expressed doubts as to whether the *zhi lan* mentioned by Confucius was a Chinese *Cymbidium*. Wei was situated in the northern part of Henan, and Lu was in Shandong, and as such, both were in the temperate regions and hence not likely to have Chinese cymbidiums growing profusely in the wild. The "*zhi lan*" could be "*ze lan*" (泽兰; *lan* of the marshes), *Eupatorium japonicum*, a fragrant herbaceous plant belonging to Asteraceae.[65,68]

The word "*lan*" also appeared in many other Chinese literary works in the same period. One example is "*Li Sao*" (离骚; *Poem of Grievance*) written by Qu Yuan (屈原; 340–278 BC), a famous statesman and poet who lived during the Warring States Period (战国时代; 476–221 BC). In it, it reads, "*I have grown nine wan (wan = 1.5–2 ha) of lan, as well as a hundred mu of hui*" (mu = 0.0667 ha).[4] However, Chen and Tang (1982) felt that it was unlikely to have such large-scale cultivation of Chinese cymbidiums during the time stated in Qu Yuan's poem. Furthermore, the description of *lan* in his poem as having green leaves and purple stems, and the whole plant, dried or fresh, being fragrant did not give a true picture of *lan*. Chinese cymbidiums have pseudobulbs (but not brown stems), and only the fresh flowers are fragrant. The *lan* mentioned in this text was most likely the fragrant grass "*lan cao*" (兰草) and not the Chinese cymbidiums. *Lan cao* was widely cultivated in China, and the whole plant was fragrant even when it was dried. During the Spring and Autumn Period and the Warring States Period, it was customary to use dried fragrant *lan cao* for bathing, sacrificial ceremony, or to kill insects.[64,65]

There were many more descriptions of *lan* in Chinese literature during the Han Dynasty (汉; 206 BC–220 AD). However, when examined closely, the *lan* mentioned in the literature is also not the same as the Chinese cymbidiums today. It was only during the Tang Dynasty (唐; 618–907) that *lan* became popular amongst the common people, and even the famous Tang poet Wang Wei (王维; 701–761) reportedly used small stones to grow *lan*.[64,65]

It is worth noting that long before Chinese *Cymbidium* was being recorded for ornamental uses, three orchids, namely, "*shi hu*" (石斛; *Dendrobium nobile*), "*tian ma*" (天麻; *Gastrodia elata*), and "*bai ji*" (白芨; *Bletilla striata*) had already been cited for use as medicine for the treatment of diseases in "*Shen Nong Ben Cao Jing*" (神农本草经; *Divine Husbandman's Classic of Materia Medica*), a pharmacological compendium published during the Han Dynasty.[65] These three orchids were also listed in "*Ben Cao Gang Mu*" (本草纲目; *The Compendium of Materia Medica*), a Chinese herbology volume written by Li Shizhen (李时珍) in 1578 during the Ming Dynasty. As of today, the three orchids mentioned above are still widely used as herbal medicine in China and Southeast Asia.[15,18,33] *Dendrobium nobile* is commonly used as a strengthening TCM tonic and is reputed to nourish the *Yin* system of the human body. It is used to alleviate thirst, calm restlessness, accelerate convalescence, and reduce dryness of the mouth. *Gastrodia elata* is an important herb used to treat headache, paralysis, epilepsy and migraine, etc. It has been considered a medicinal herb of choice in the battle against dementia. *Bletilla striata* is known for treating internal bleeding, tuberculosis, gastric and duodenal ulcers. Additional applications include protection against flatulence and dysentery.[15,18,23,33]

1.3. Literature on Cultivation of Chinese *Cymbidium*

According to Chen and Ji,[65] the earliest accurate record of involving Chinese cymbidiums as *lan* in Chinese literary work is "*Yong Lan*" (咏兰) by Tang Yanqian (唐彦谦), published in the late Tang Dynasty about 900 AD. The description of the leaves and flowers of the *lan* was consistent with that of Chinese cymbidiums today. Another credible piece of literature on the cultivation of Chinese cymbidiums published in the late Tang Dynasty is "*Zhi Lan Shuo*" (植兰说; *Growing Lan*) by Yang Kui (杨夔), and it is one of the earliest examples of literature that provides a good description of the cultivation of *lan*.

Chinese cymbidium cultivation became widespread during the Song Dynasty (宋; 960–1279), and literature on the descriptions of characteristics, ecology, and distribution of Chinese cymbidiums became more abundant. Worthy of being mentioned here is the work of Huang Tingjian (黄庭坚; 1045–1105). He was a poet and a calligrapher, as well as a keen Chinese orchid lover in the Northern Song Dynasty (北宋; 960–1127). He was credited with making the distinction between "*lan*" (兰) and "*hui*" (蕙).[47,64] In his book, "*You Fang Ting*" (幽芳亭), he wrote:

一干一华而香有余者兰，
一干五七华而香不足者蕙。

Translated, it means

A flowering stalk with one single fragrant flower is lan;
A flowering stalk with 5–7 but less fragrant flowers is hui.

Although Chinese *Cymbidium* cultivation was common during the Song Dynasty, it became even more popular during the Ming (明; 1368–1644) and Qing (清; 1644–1911) Dynasties. These are two very important periods in the history of the development of Chinese orchid culture. Much of this popularity can be attributed to the increasing importance of the development of light industries.

There were an ever-growing number of businessmen, bureaucrats, and scholars, and they brought about a renewed interest in orchid cultivation. Selected early literature on the cultivation of Chinese cymbidiums appeared during the Song, Ming and Qing Dynasties (1233–1876) are listed in Table 1.1.

Table 1.1. Selected early literature on Chinese *Cymbidium* cultivation during the Song, Ming and Qing Dynasties (1233–1876).

Period	Title	Author	Year of Publication
Song Dynasty	金漳兰谱 Jin Zhang Lan Pu	赵时庚 Zhao Shigeng	1233
	王氏兰谱 Wang Shi Lan Pu	王贵学 Wang Guixue	1247
	兰谱奥法 Lan Pu Ao Fa	赵时庚* Zhao Shigeng	
	兰易 Lan I	鹿亭翁* Lu Tingweng	
Ming Dynasty	兰史 Lan Shi	簟溪子* Dian Xizi	
	本草纲目 Ben Cao Gang Mu	李时珍 Li Shizhen	1578
	遵守八笺兰谱 Zun Shou Ba Jian Lan Pu	高濂 Gao Lian	1591
	群芳谱 Qun Fang Pu	王象晋 Wang Xiangjin	1621
Qing Dynasty	第一香笔记 Di Yi Xiang Bi Ji	朱克柔 Zhu Kerou	1796
	兰蕙镜 Lan Hui Jing	屠用宁 Tu Yongning	1811
	兴兰谱略 Xing Lan Pu Lue	张光照 Zhang Guangzhao	1816
	兰言述略 Lan Yan Shu Lue	袁世俊 Yuan Shijun	1876

Note: *Please refer to text (Chapter 1).
Source: Refs. [12,47,55,65,68].

The most famous earliest publication that provided detailed information on Chinese *Cymbidium* cultivation is *Jin Zhang Lan Pu* (金漳兰谱; *Treatise of Orchids of Jin Zhang*). It was written by Zhao Shigeng (赵时庚) and was published in 1233 during the Southern Song Period (Fig. 1.2). It is also believed to be the earliest book on orchids in the world.[65] The publication is divided into five chapters, covering 32 different types of *lan* species, mainly from Fujian. They are divided into two groups: "*Zi Lan*" (紫兰; *Violet Lan*) and "*Bai Lan*" (白兰; *White Lan*). This publication describes in fair details the

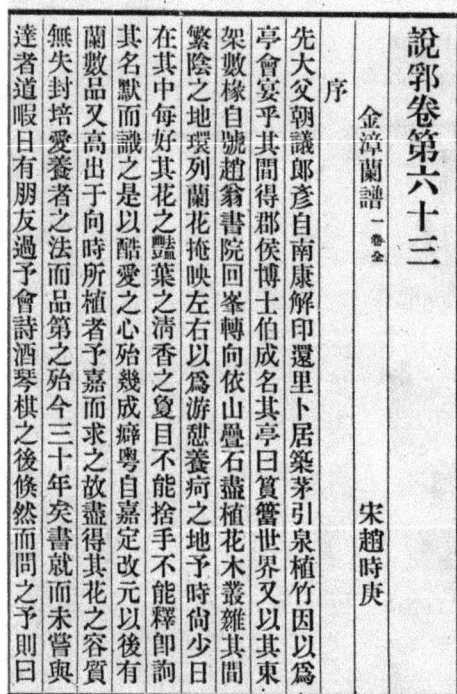

Fig. 1.2. A reprint of "*Jin Zhang Lan Pu*" (金漳兰谱) written by Zhao Shigeng (赵时庚), a famous book on Chinese *Cymbidium* cultivation published in 1233 during the Southern Song Dynasty.

morphological characteristics of the two groups of *lan* and provides good information regarding their cultivation, irrigation, transplantation, soil quality and fertiliser application.[47,60]

In 1247, a book entitled "*Wang Shi Lan Pu*" (王氏兰谱; *Wang's Treatise on Orchids*) on the cultivation of Chinese *Cymbidium* written by Wang Guixue (王贵学), was published in the Southern Song Dynasty (Fig. 1.3). It was an improved version of *Jin Zhang Lan Pu* and provided a good description of the types of *lan*, watering techniques, division methods, the use of soil or soil mixtures as potting media, and cultivation management.[65]

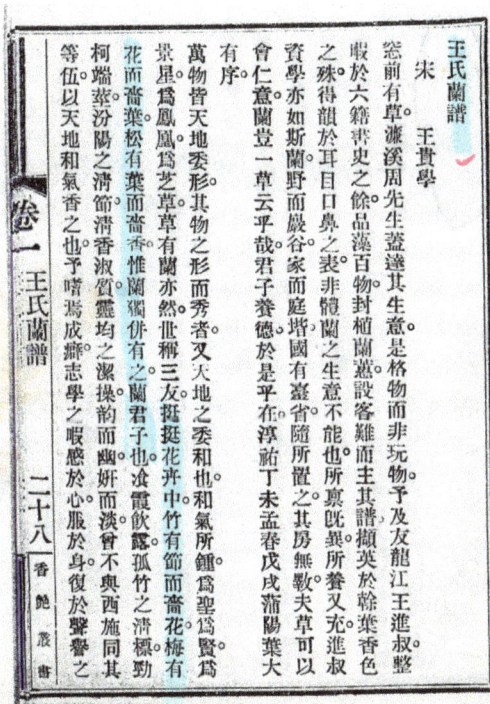

Fig. 1.3. A reprint of "*Wang Shi Lan Pu*" (王氏兰谱) written by Wang Guixue (王贵学), another famous publication of Chinese *Cymbidium* cultivation published in 1247 during the Southern Song Dynasty.

Another publication on Chinese *Cymbidium* cultivation published during the Song Dynasty was "*Lan Pu Ao Fa*" (兰谱奥法; *Treatise on the Technique of Orchid Culture*). It is generally believed that it was written by Zhao Shigeng (赵时庚), and Zhou Lv Qing (周履清) was involved in the editing of the publication.[47,55,65] It provides the prevailing cultural practices of orchid cultivation in greater details. This includes propagation by division, repotting, use of soil collected from their habitats, frequency of watering, fertiliser application and pest control.[60,65]

Two other publications are worthy of mention:

1. "*Lan I*" (兰易; Parts I and II) — "*I*" (易) in Taoism means "*change*" as in "*I Ching*" (易经; *The Book of Change*)
2. "*Lan Shi*" (兰史; *A History of Lan*).

Early literature indicated that *Lan I* was written by Lu Tingweng (鹿亭翁), whereas *Lan Shi* was written by Dian Xizi (簟溪子). However, according to the research done by Yu Jiaxi (余嘉锡; 1884–1955), *Lan I* and *Lan Shi* were the works of Feng Jingdi (冯京第) during the Ming Dynasty.[47,65] Cultivation practices of *Cymbidium* orchids were divided into 12 months based on the "*Ba Gua*" (八卦), the eight divinatory trigrams of Taoist practices. It describes the temperature suitable for orchid growth, how to water and fertilise the plants and increase soil fertility, and what methods could be used to drive away harmful insects, according to each month of the year.[19]

During the Ming Dynasty, the most celebrated publication was "*Ben Cao Gang Mu*" written by Li Shizhen in 1578. The book lists plants, animals, and minerals that have medicinal properties. More importantly, it also has a relatively complete description of the names, categories and medicinal uses of the Chinese orchids. It also clearly discussed the differences between some common fragrant plants, such as *Eupatorium japonicum* and Chinese *Cymbidium*. He also pointed out the major morphological differences between *C. goeringii* and *C. ensifolium*.[65]

Another notable work, "*Di Yi Xiang Bi Ji*" (第一香笔记), written by Zhu Kerou (朱克柔), was published during the Qing Dynasty in 1796. This book consists of four volumes and is divided into seven sections that provide valuable information on flower morphology, cultivation, protection, miscellaneous notes, citations, and others. In the chapter on flowers, the morphology and flowering characteristics were carefully studied and described. Many technical terms created in the book are still being used today.[47]

Most of the earlier publications on the horticultural practices of Chinese cymbidiums are rather brief. However, *Di Yi Xiang Bi Ji* provides us with substantial useful information regarding the classification, morphology, history and horticultural practices of Chinese cymbidiums.

There was a surge of publications on orchid cultivation in China in the 20th century. "*Lan Hui Xiao Shi*" (兰蕙小史) compiled by Wu Enyuan (吴恩元), and published in 1923, further summarised the traditional varieties of Chinese cymbidiums at that time. The book consists of three volumes focusing on the 161 popular Chinese cymbidiums found in the Zhejiang region. It systematically recorded the various types of floral form and cultivation management. It also included sketches of many famous lost species of Chinese cymbidiums and beautiful illustrations of the species (Fig. 1.4), making it the first comprehensive early book on *lan* with pictures.[60]

With the rapid increase of Chinese *Cymbidium* varieties and better experience in cultivation management, orchid cultivation became popular amongst the Chinese public. Towards the late 20th century, publications based on scientific research started to appear. Suffice to say, books on Chinese *Cymbidium* published in the late 20th century are much better and detailed in terms of photographic quality and scientific contents. Some selected publications include the following:

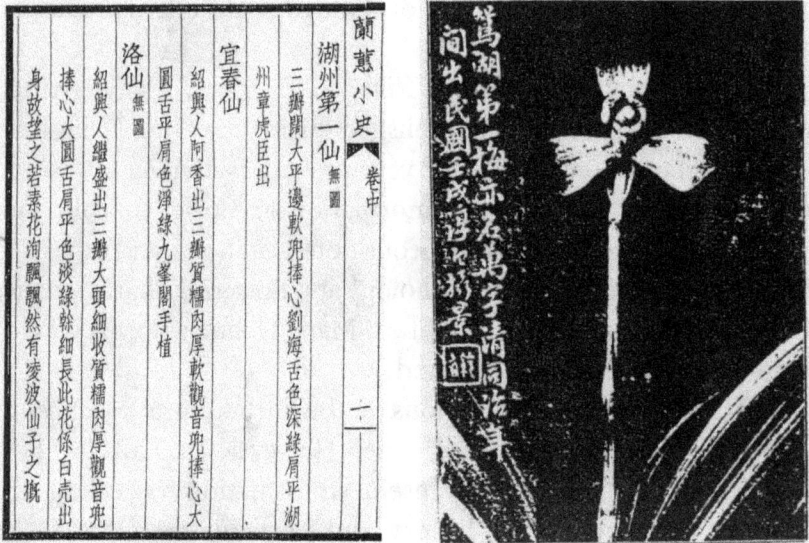

Fig. 1.4.   A reprint of *"Lan Hui Xiao Shi"* (兰蕙小史) compiled by Wu Enyuan (吴恩元), who described some of the traditional varieties of Chinese *Cymbidium* published in 1923. It is the first book to include sketches of many lost famous species of Chinese cymbidiums.

- *Lan Hua* (兰花; 1980) and *Zhong Guo Lan Hua* (中国兰花; 1991) by Wu Yingxiang (吴应祥).
- *Zhong Guo Lan Hua* (中国兰花; 1990), edited by He Qingzheng and Chen Xinqi (何清正，陈心启).
- *Lan Hua Zai Pei Ru Men* (兰花栽培入门；1990）by Lu Sicong (卢思聪).
- *Lan Hua* (兰花; 1991) by Liu Qingyong (刘清涌).
- *Zhong Guo Lan Hua* (中国兰花; 1995) by Li Shaoqiu, Hu Songhua, and Lu Zhang (李少球, 胡松华, 鲁章).
- *Zhong Guo Lan Hua Quan Shu* (中国兰花全书; 1997) by Chen Xinqi and Ji Zhanhe (陈心启, 吉占和).
- *Zhong Guo Lan Zhu Zhi Wu* (中国兰属植物; 2006) by Liu Zhongjian, Chen Xinqi, and Ru Zhengzhong (刘仲健, 陈心启, 茹正忠).

- *Guo Lan Sheng Li* (国兰生理; 2006) by Pan Ruichi and Ye Qingsheng (潘瑞炽, 叶庆生).
- *Zhong Guo Lan Wen Hua* (中国兰文化; 2008) by Ma Xingyuan and Ma Yangchen (马性远, 马扬尘).
- *Guo Lan Ming Pin Shang Jian* (国兰名品鉴赏; 2011), edited by Xu Dongsheng (许东生).
- *Lan Hun* (兰魂; 2016) by Wang Jiaxing (王家兴).
- *Zhong Guo Lan Hua Ming Pin Zhen Pin Jian Shang Tu Dian* (中国兰花名品珍品鉴赏图典; 3rd ed., 2020) by Liu Qingyong (刘清涌), and many others.

Chapter 2

Chinese *Cymbidium* and Chinese Culture

Over thousands of years, the life and culture of the Chinese have long been associated with Confucianism, Buddhism, and Taoism, which have shaped the Chinese culture and become an integral part of Chinese civilisation. The influence of these three philosophies on Chinese history, literature, philosophy, religion, politics, science, arts, medicine and others has remained very strong. The association of Chinese *Cymbidium* to Chinese culture is, therefore, no exception.

2.1. The *Lan* Culture

Cultivation of Chinese *Cymbidium* has a long history in China. Naturally, the aesthetic principles, appreciation and judging criteria of *lan* are greatly influenced by the practices and cultural beliefs in China. Traditionally, the Chinese aesthetic concept is based on the model of a righteous, noble, pure and simple character, which conforms to the teachings of Confucianism, Buddhism, and Taoism, and also to the humble mentality of the Chinese people.[76,79,89] The impact of the three great Eastern philosophies on Chinese *Cymbidium* culture is best summarised by Liu,[64] who highlighted that Confucianism *"respects lan"* while Buddhism and Taoism *"love lan"*.

It is a common practice in Confucius's philosophy to personify plants, animals, mountains and other living things. As mentioned before, Confucius introduced Chinese *Cymbidium* into Chinese culture, using it as a kind of cultural image to build up the concept of a spiritual ideal that people need to pursue. Confucius's teaching focuses on the development of human noble characters, emphasises the importance of personal ethics and social morality, and also explains what the characteristics of a gentleman are. Confucianism *respects lan* as it symbolises the integrity and nobility of a gentleman, qualities that are worth admiring.[89]

The impact of Confucianism on the Chinese *Cymbidium* culture is evident in the criteria used in the appreciation and judging of Chinese cymbidiums. It corresponds well with the moral standards of Confucianism, such as "*The Doctrine of the Mean*" (中庸) and "*Modesty*" (谦让). For example, when judging Chinese cymbidiums, the flowers cannot be without fragrance or emit a strong fragrance. A moderate, pure and delicate fragrance is the best. The Chinese are also very particular about the colour of the flowers. If the flower is too dull or pale in colour, it appears lifeless, but at the same time, an extremely striking and bright colour makes one feel overwhelmed. Therefore, colours like white, green, or light yellow are most loved by *lan* lovers, tying in with the Chinese values of harmony and moderation.[47]

Buddhists and Taoists love *lan* as it is simple, humble and pure. Buddhism teaches that all living things are part of nature and, hence, should co-exist in harmony and be treated with love. It also teaches that by facing the reality of life positively, one can achieve a simple and happy life. Taoism stresses the harmony between man and nature and the balance between the two opposite forces — *Yin* and *Yang*. It emphasises the importance of going along with the flow of nature and not to go against it. Tying into this, *lan* can grow in very poor-quality soil on mountain cliffs and is also able to grow among wild bushes. Its unique character happens to have something in common with the ideal state pursued by both Buddhists and Taoists.

Harmonious co-existence and humility have profound meanings in Chinese traditional culture.[76,79,86,89]

It is not surprising that the names of many varieties of Chinese cymbidiums often include terms associated with Confucianism, Buddhism and Taoism. For instance, we have names of Chinese *Cymbidium* varieties relating to **Confucianism**, such as 永福 (blessing to be happy forever) and 瑞雪丰年 (timely and auspicious snow bringing in a year of abundance); relating to **Buddhism**, such as 佛之光 (Buddha's light; the light radiated from the bodhisattva's head wheel), 圣佛 (Holy Buddha), and 菩提 (Bodhi means supreme enlightenment); and relating to **Taoism**, such as 仙殿 (Immortal palace) and 天鹤 (crane; Taoism recognises cranes as the incarnation of immortals). This shows how Chinese *Cymbidium* culture incorporates the connotations of these three great Eastern philosophies.[47,64]

According to records,[67] the close relationship between *lan* and Buddhist culture began in the Tang Dynasty. It was Guan Xiu (贯休; 832–912), a celebrated Buddhist monk, painter, poet and calligrapher living in Lanxi (兰溪), Zhejiang province, who started to grow and write about *lan*. Since then, it has become a tradition to grow *lan* in most of the temples. As both Buddhists and Taoists seek a pure and simple life, and they find peace and tranquillity in *lan*, it is common to have *lan* as a companion when practising meditation. The calm and sweet-smelling atmosphere induces people to feel refreshed and relaxed, and hence, *lan* is also known as the "*Zen Flower*" (禅花).

Buddhists and Taoists are also very particular as to where they build their temples. The ideal and proper geographical location and environment are of great importance. In China, it is quite common to see temples on top of a mountain or in forests. Under the influence of Tao values and practice, appreciation and love for nature became increasingly popular amongst the Chinese people. It became a popular practice for scholars, government officials, poets, and artists to retreat and live in a secluded place in the mountains amidst a beautiful landscape and environment. This was particularly so for senior government officials who were frustrated with politics and corruption or those who had fallen out of favour. Temples in mountains were the most preferred retreat places as the surrounding air was fresh and pure. Together with moderate rainfall and sunshine, such an environment also favoured the growth of Chinese cymbidiums in the wild. It is well documented that Chinese cymbidiums grow well in mountains and valleys. This is the natural habitat where one is likely to find plentiful wild orchids.

In Sichuan, for example, there are two famous mountains, Mount Emei (峨眉山) (Fig. 2.1) and Mount Qingcheng (青城山) (Fig. 2.2). While both are known for their beautiful mountainous landscape and peaceful atmosphere, attracting many tourists, they are also known to produce some of the best Chinese *Cymbidium* species.[60,64]

Mount Emei, one of the Four Sacred Buddhist Mountains, commonly known as "*The bodhimandala of Samantabhadra Bodhisattva*" (普贤菩萨道场) is 3,099 m above sea level. The interplay of its topography, climate conditions, soil structure and vegetation makes it a unique ecological system well suited for orchid

Fig. 2.1.   Mount Emei (峨眉山), one of the Four Sacred Buddhist Mountains in China.

Fig. 2.2.   Mount Qingcheng (青城山), the birthplace of Taoism in China.

Fig. 2.3.   The Buddha Orchids (佛兰).[66]

growth, i.e., high relative humidity, cool weather and moderate sunlight. It produces some of the best *C. goeringii* and *C. kanran*. An exquisite Chinese orchid known as "*Buddha Orchid*" (佛兰) (Fig. 2.3) was collected at Mount Emei in 1988, and the structure of the pollen cap of this special orchid resembled Buddha in meditation, hence earning its name.[60,66] In 2005, one of the exotic Chinese *Cymbidium* varieties "*Oriental Buddha Orchid*" (东方佛兰) was reported to be auctioned for over ten thousand RMB.

Mount Qingcheng is the birthplace of Taoism[60] and has become one of the most significant Taoist centres. It is surrounded by mountains, deep valleys and forests and experiences heavy rainfall. Mount Qingcheng is known as the natural treasure house for wild orchids in China because of its rich orchid flora. Many of the award-winning Chinese *Cymbidium* species also originated from here; for example, in 1992, the famous *Buddha Orchid* was also discovered in Mount Qingcheng.[60] In Taoist scriptures, *lan* is regarded as "*immortal grass*" (仙草), and it represents a holy and perfect temperament. Chinese cymbidiums have been widely planted in Taoist temples since the Tang Dynasty. Temples on Mount Qingcheng in Sichuan continue to keep the tradition of planting Chinese orchids alive even today.

2.2. *Lan* in Ancient Chinese Literature

Lan has been a favourite historical subject for creative writers in Chinese literature. As mentioned earlier, one of the important factors that contributed to the extensive use of *lan* in Chinese literature was the influence of Confucius. He likened himself to a *lan* growing in the wild but not being appreciated. Many frustrated brilliant Chinese scholars at that period also had similar sentiments.[19]

One of the earlier examples that reflects this sentiment is found in a poem entitled "*Grievance*" (怨篇), written by Zhang Heng (张衡; 78–139), a great astronomer in the Eastern Han Dynasty (东汉; 25–220):

猗猗秋兰，植彼中阿。
有馥其芳，有黄其葩。
虽曰幽深，厥美弥嘉。
之子云远，我劳如何！

Hu[19] has translated it as follows:

"*How elegant, O orchid of autumn!*
The sweet smell, the yellow bud.

Though confined in shade; Your beauty is beyond comparison.
As afar as the cloud. I suffer; O what!"

The poet first described the Chinese *Cymbidium* to be so still yet beautiful and bursting with fragrance despite growing in a deep mountainous area. However, he felt sad that the orchid he appreciated was so far away in the wild. He felt helpless, and this reflected his feelings and sentiment of being unrecognised and unappreciated for his talent, just like the orchid, which is so fragrant yet not sought after. The author used the orchid as a metaphor for the lack of appreciation of his talent during that period and as a reflection of his disheartened acceptance of his destiny and misfortune in not being recognised.

Another good example that describes the same sentiment is the poem "*Drinking Wine*" (饮酒) written by Tao Yuanming (陶渊明; 365–427), a famous Chinese poet born during the Eastern Jin Dynasty:

幽兰生前庭，含薰待清风。
清风脱然至，见别萧艾中。

Hu[19] has translated it as follows:

"*The orchid in the frontcourt,*
Holding her perfume to wait for the morning breeze,
The breeze passes unnoticed,
Clumps of Hsiao-ai distinguished."

Two terms appear in this poem, namely, "*hsiao-ai*" (萧艾; *Artemisia*) and "*qing feng*" (清风; cool and pure breeze). *Hsiao-ai* is a persistent, unpleasant weed growing in the wild that aggressively deprives other plants of nutrients. It has, therefore, been referred to as "*xiao ren*" (小人; a person of low social status or an unscrupulous man). This term is used in contrast to "*jun zi*" (君子; the perfect gentleman), which is symbolised by *lan*. The cool breeze, *qing feng*, will bring out the fragrance of *jun zi*, distinguishing *lan* from the *hsiao-ai*.

When the terms "*qing feng*" and "*jun zi*" are used together, it would mean you are looking for an opportunity to meet the right person with authority who appreciates your talent and integrity. This would, however, depend on how good the authority or government was at the time. Tao Yuanming, for example, was very sceptical and had strong qualms that the government was fair and caring. He was a government official before and was frustrated with the politics in the government and had opted for early retirement to lead a simple countryside life.

Chen Yi (陈毅; 1901–1972), a marshal and politician in modern China, was more positive and forward-looking, as evident in his poem "*Yulan*" (幽兰).[47]

幽兰生山谷, 本自无人识,
只为馨香重, 求者遍山隅。

It means the following:

"*Yulan grows in the deep valley, but no one notices it.*
Only because of its fragrance, people will search for
it all over the mountains".

He believed that if one was talented and had a lofty character, he would be well sought after no matter where he was. In ancient China, a talented person living in a secluded place in the mountain or in temples leading the life of a hermit was often referred to as a *lan* growing in the wild.

2.3. *Lan* in Ancient Chinese Paintings and Calligraphies

Painting is one of the more common art forms where *lan* is associated with Chinese culture. In the olden days, Chinese paintings were mostly painted on silk. However, after paper was invented, silk was gradually phased out. Chinese *Cymbidium* paintings are generally drawn in black and white, with very few artists using other colours. Plain colour is preferred because of the Chinese belief that "*su-ya*" (素雅; simplicity and elegance) is the unique characteristic of the Chinese *Cymbidium*. Hence, black and white are the most effective media to express this quality.[19] Chinese orchid paintings are remarkably simple in composition. Simplicity is the underlying philosophy of Chinese orchid paintings.

Paintings of Chinese *Cymbidium* emphasise the overall beauty of the whole plant. Orchid painters appreciate not only the elegant and beautiful flowers but also focus on the beauty of the leaves. As Chinese people always say, "观花一时，赏叶经年", which means one looks at the flowers for merely a moment but will admire the leaves for years to come.

Famous painters, such as Zhao Mengjian (赵孟坚; 1199–1267) in the early Song Dynasty, Xu Wei (徐渭; 1521–1593) (Ming Dynasty) and Zheng Banqiao (郑板桥; 1693–1765) in the Qing Dynasty were very particular about painting the Chinese *Cymbidium* leaves. They spent significant efforts in presenting a good illustration when highlighting the features and beauty of the Chinese orchid leaves. This shows that the leaves are an important part of the overall beauty and artistic concept of the Chinese *Cymbidium*. It is only when the flowers and leaves are both nicely painted that it can then be regarded as an excellent Chinese *Cymbidium* painting.[60,64,87]

For a beginner in Chinese *Cymbidium* painting, it is recommended that one observes the growth characteristics and understands the ecological habits of the plants. Beginners should also study the works of grandmasters and learn how the elegant and virtuous spirit of orchids is reflected in their works. Of all the popular plants in China, such as *plum blossom, Chinese Cymbidium (lan), bamboo, and chrysanthemum,* painting *lan* is by far the most difficult. It may take years of practice before one reaches a certain level of proficiency. There is an old Chinese saying "十年画竹, 百年画兰", which means "*The painting of bamboo takes ten years while the painting of lan takes a hundred*".[49]

Early Chinese Cymbidium paintings

According to Chen and Tsi,[65] the earliest Chinese *Cymbidium* painting currently preserved in the Palace Museum in Beijing is a traditional watercolour painting of *C. faberi* on a silk fan by a court painter from the Northern Song Dynasty (Fig. 2.4). Another famous painting, "*Spring Orchid*" (春兰图) by Zhao Mengjian of the Southern Song Dynasty, regarded as the earliest Chinese orchid painting in existence, is also kept in this museum. He was the first person to use Chinese black ink to paint Chinese orchids. He painted two *C. goeringii*, growing in grasslands, with freshly opened flowers dancing like butterflies[60] (Fig. 2.5). Zhao was the 11th generation descendent of Zhao Kuangyin (赵匡胤; 927–976), the founding emperor of the Song Dynasty. He was good at poetry, calligraphy and Chinese orchid painting. After the fall of the Southern Song Dynasty, he left the government services and lived in seclusion in Jiahe (嘉禾), Hunan.[64,65]

Fig. 2.4.　The meticulous watercolour painting of *C. faberi* on a silk fan by a court painter from the Northern Song Dynasty (960–1126). (This painting is presently conserved in The Palace Museum in Beijing).

Fig. 2.5. *"Spring Orchid"* (*C. goeringii*) (春兰图), painted by Zhao Mengjian (赵孟坚) (1199–1267) during the Southern Song Dynasty. (This is the earliest famous Chinese *Cymbidium* painting currently conserved in The Palace Museum in Beijing).

Another great work, the *"Ink Orchid"* (墨兰图), is a painting by Zheng Sixiao (郑思肖; 1241–1318), who was born in Lianjiang (连江), Fujian province. When the Southern Song Dynasty was conquered by the Mongolians in 1279, he refused to serve the Yuan government and moved to Wu Xia (吴下; now known as Suzhou) and led the life of a hermit. Being a patriot, he always sat facing the south where the Southern Song was situated and refused to face the north where the Yuan capital was located. He even renamed himself *"Suo Nan Weng"* (所南翁), which means "Old man sitting facing south", and that name appears in many of his paintings.[65]

Zheng Sixiao was a man of strong principles. When he was living in seclusion in Jiangnan, a county magistrate asked Zheng to do a painting of *C. sinense*, but he refused. The county magistrate threatened to put him in jail. Zheng furiously said, "You can chop off my head, but you cannot have my orchid painting."[47] His famous painting, *"Ink Orchid"*, was "Orchids without root and soil" (Fig. 2.6). This was symbolic as Zheng was trying to convey the message that his motherland originally under the Song Dynasty was occupied by the Yuan Dynasty, and was like an orchid growing without soil, reflecting his feelings of being groundless and stateless. The rootless and soilless orchid painting

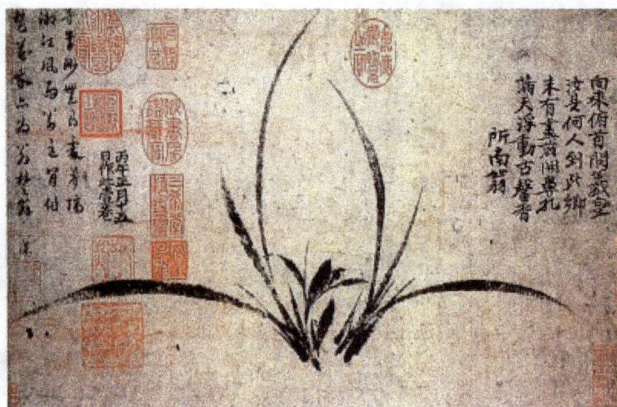

Fig. 2.6.   "*Ink orchid*" (*C. sinense*) (墨兰图), an ancient painting by Zheng Sixiao (郑思肖) (1241–1318) during the Southern Song Dynasty. (This painting is presently conserved in the Osaka City Museum of Fine Arts).

was representative of the strong nationalistic feelings and pain he experienced upon seeing his homeland being taken over by invaders. Since then, *lan* became a symbol of loyalty. This painting is now kept in the Osaka City Museum of Fine Arts, Japan.[47,65]

There was a rapid increase in Chinese orchid paintings after the Yuan Dynasty. Many renowned orchid painters appeared during the Ming Dynasty (1368–1644), Qing Dynasty (1644–1911), and the early Republic of China. It has been estimated that there are collections of 33 orchid paintings by at least 11 painters during the Ming period, and 101 collections by 32 painters during the Qing period, all being kept in museums around the world.[64,65] Table 2.1 shows some of the better-known Chinese *Cymbidium* painters in China from 1097 to 1971.[47,50] Selected paintings by some of the painters in this list are shown in Figs. 2.7–2.16. Five well-known orchid painters on the list — Wang Shishen, Li Shan, Luo Pin, Zheng Xie, and Li Fangying — were amongst the "*Eight Eccentrics of Yangzhou*".

"*Eight Eccentrics of Yangzhou*" (杨州八怪)

"*Eight Eccentrics of Yangzhou*" generally refers to eight of the most prominent Chinese *Cymbidium* painters and calligraphers. They were active in Yangzhou during the Qing Dynasty.[8] However, there are disagreements by some orchid experts or historians as to who these eight persons were. The generally accepted list of the *Eight Eccentrics of Yangzhou* is as follows:[75,83]

Li Shan (李鱓) (1686–1756)
Wang Shishen (汪士慎) (1686–1759)
Jin Nong (金农) (1687–1763)
Huang Shen (黄慎) (1687–1772)
Gao Xiang (高翔) (1688–1753)
Zheng Xie (郑燮) (1693–1765)
Li Fangying (李方膺) (1695–1754)
Luo Pin (罗聘) (1733–1799)

Table 2.1. List of some prominent painters of Chinese *Cymbidium* in China (1097–1971).

Yang Wujiu (杨无咎) (1097–1171)
Zhao Mengjian (赵孟坚) (1199–1267)
Zheng Sixiao (郑思肖) (1241–1318)
Zhao Mengfu (赵孟頫) (1254–1322)
Puming (普明) (unknown–1352)
Wen Zhengming (文征明) (1470–1559)
Xu Wei (徐渭) (1521–1593)
Feng Zhaoqi (冯肇杞) (1612–1670)
Shi Tao (石涛) (1642–1708)
Li Shan (李鱓) (1686–1756)
Wang Shishen (汪士慎) (1686–1759)
Zheng Banqiao (郑板桥) (1693–1765)
Li Fangying (李方膺) (1695–1754)
Luo Pin (罗聘) (1733–1799)
Wu Changshuo (吴昌硕) (1844–1927)
Qi Baishi (齐白石) (1863–1957)
Pan Tianshou (潘天寿) (1897–1971)

Source: Refs. [47,50,60].

Fig. 2.7.   Painting by Wen Zhengming (文征明) (1470–1559). (The painting is presently conserved in The Palace Museum in Beijing).

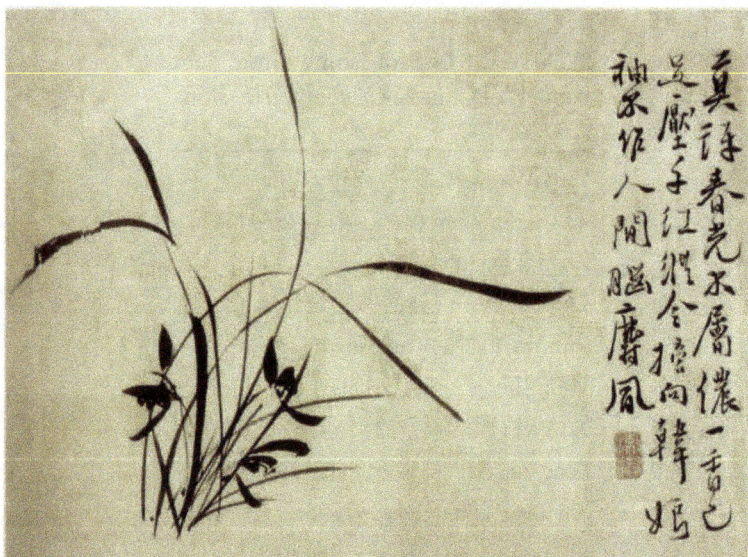

Fig. 2.8.   Painting by Xu Wei (徐渭) (1521–1593).

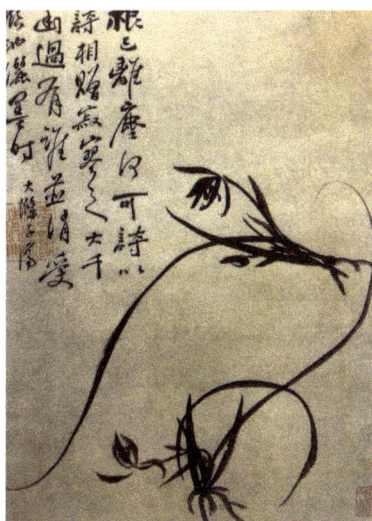

Fig. 2.9.   Painting by Shi Tao (石涛) (1642–1708).

Fig. 2.10.   Painting by Li Shan (李鱓) (1686–1756).

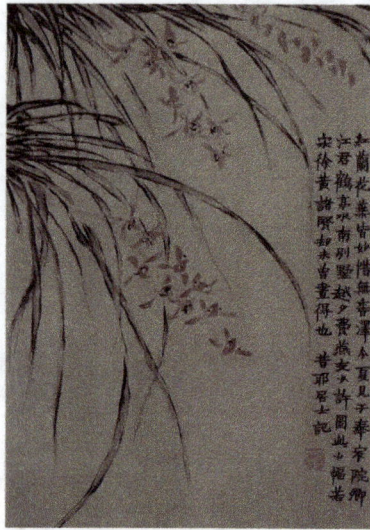

Fig. 2.11.   Painting by Jin Nong (金农) (1687–1763).

Fig. 2.12.   Painting by Li Fangying (李方膺) (1695–1754).

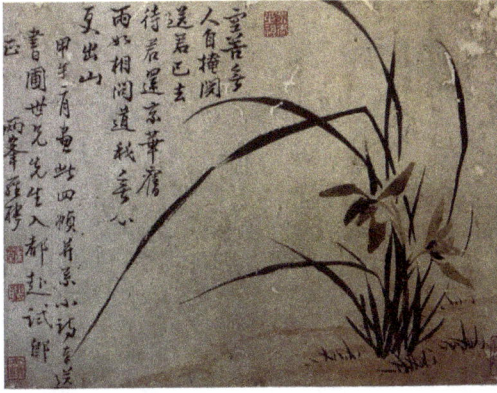

Fig. 2.13. Painting by Luo Pin (罗聘) (1733–1799).

Fig. 2.14. Painting by Wu Changshuo (吴昌硕) (1844–1927).

Fig. 2.15.   Painting by Qi Baishi (齐白石) (1863–1957).

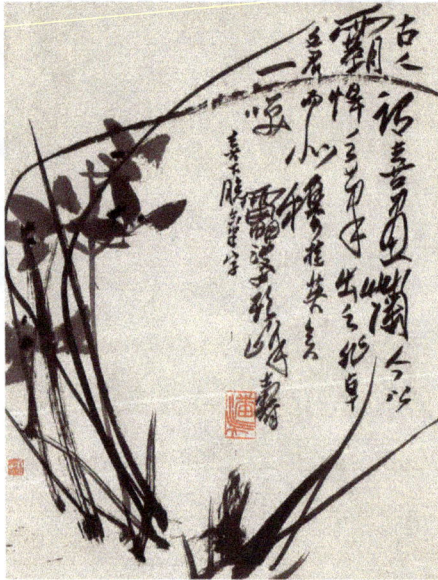

Fig. 2.16.   Painting by Pan Tianshou (潘天寿) (1897–1971).

The term "*Eight Eccentrics of Yangzhou*" was more of a statement about their artistic style rather than any social eccentricities. Most of them were from impoverished backgrounds and were disgusted by the corrupted and despicable behaviours that prevailed in the government circles back then. They rejected rigid and bureaucratic ideas and pursued what they felt was a realistic view of the daily life of the masses in their calligraphy and painting. This often exposed the ugly side of the society and high officials, making the artists very unpopular with the ruling class. For these reasons, they were sometimes branded as the "*Eight Monsters of Yangzhou*".

Each of the eight eccentrics had his unique painting style, reflecting the painter's character and background. For instance, the Chinese *Cymbidium* painted by Zheng Banqiao is elegant and simple, illustrating how the orchid grows in the mountains in harmony with nature. His paintings give people a sense of serenity (Fig. 2.21–2.23).

In comparison, the Chinese *Cymbidium* painted by Li Fangying is wild, vigorous, unrestrained — leaves fluttering and not following the usual style (Fig. 2.12). This was probably related to his unfortunate personal experiences as well as his bold, unfettered and unrestrained character. According to historical records, Li Fangying was born in a family of officials. He served as a county magistrate for many years. He was honest and caring towards people but was unfortunately dismissed from the imperial court due to false accusations. He later made a living by selling paintings. He often used paintings and poems to express his loneliness, dissatisfaction and frustration and felt that his talent was not appreciated.[50,70,77]

Li Shan, another of the eight eccentrics, was a native of Xinghua, Jiangsu. He was an imperial court painter for 53 years but was forced to leave because he did not want to be bound by the rigid and constrained style of painting. After being demoted, he went to Yangzhou to sell paintings for a living. In Yangzhou, he was inspired by the brushwork of Shi Tao (石涛), one of the early famous

Fig. 2.17.   A sculpture of the "*Eight Eccentrics of Yangzhou*" displayed at the Memorial Hall in Yangzhou.

painters, and started developing his unique style of painting. Li's painting style is bold and unrestrained, with vigorous and free strokes[50] (Fig. 2.10). In 1997, the Ministry of Posts and Telecommunications of the People's Republic of China issued a New Year's postcard with the themes of plum, bamboo, Chinese *Cymbidium,* and chrysanthemum. The Chinese *Cymbidium* illustration was derived from one of the paintings of Li Shan.[47]

The style of paintings pioneered by the eight eccentrics emphasised nature and the daily life of people. Today, thousands of paintings produced by the eight eccentrics are kept in over 200 art galleries and research institutes in China and overseas.[62] A museum dedicated to the memory of the eight eccentrics was also established in Yangzhou (Fig. 2.17).

Zheng Banqiao (郑板桥; 1693–1765)

The most famous of the *Eight Eccentrics of Yangzhou* was Zheng Xie (郑燮), popularly known as Zheng Banqiao. He was the greatest painter of Chinese *Cymbidium* and bamboo.

Zheng came from a learned but poor family in Jiangsu. He was the only child, and his mother died when he was three years old. At the age of 19, he was a "*xiu cai*" (秀才) in the county level examination. When he was 40, Zheng was a "*Ju ren*" (举人) in the imperial provincial examination. At the age of 44, he became a "*jin shi*" (进士) in the highest level of imperial examination. Zheng started to work as a district magistrate when he was 50 years old, first in Fan County (范县) and later in Wei County (潍县). He was very well respected because of his concern for the poor and the sick.

However, his caring and humanitarian approach made him very unpopular with the corrupt high officials and unscrupulous businessmen who did not see the need to care for the public. It was reported that in one year, when Wei County (which was under his care) was experiencing a very severe famine, he decided to open the government grain depot for the starving masses without obtaining prior approval from the central government, and because of that, he was relieved of his post. He then decided to devote all his time to painting in Yangzhou.[81]

Zhang's care for the poor and his dislike for the unscrupulous businessmen and the corrupt bureaucrats were reflected in many of his poems. In one of his poems, he wrote the following[47]:

凡吾画兰、画竹、画石，
用以慰天下之劳人，
非以供天下之安享人也。

Translated literally, it means, "*My paintings of lan, bamboo, and boulder are meant to comfort the hardworking masses, not to please those who lead a comfortable and leisurely life*".

Zheng respected and loved to paint *lan*, bamboo, and boulders with good reasons. He once said that *lan* bloomed in all seasons, bamboo was evergreen with joints (meaning *integrity*), and boulders remained unchanged for hundreds of years. He treated them as sacred objects. In many of his paintings, you will find that the *lan* grows singularly or in groups in the cracks of boulders, on cliffs, or on mountainous slopes, interspersed with bamboos amongst boulders — all living harmoniously. His paintings bring out the integrity and unyielding unique characters of *lan* and bamboo.[47]

Zheng Banqiao was praised for being an honest and upright officer. When his daughter was getting married, he could only give her one of his Chinese *Cymbidium* paintings as dowry as he did not have much savings. There are many stories that describe Zheng's concern and compassion for the poor working class when he was a district magistrate. One of the stories depicts his caring attitude towards the poor and people in need. In the story, Zheng was on his way to a gathering. Along the way, he saw a poor old lady selling foldable bamboo fans. However, nobody was buying one from her. Noticing that the fans were only made with a white sheet of paper, he hurried to a shop nearby, borrowed a Chinese painting brush, ink, and ink slab, and began helping the old lady by painting *lan* and bamboo on the fans. The passers-by saw the work from the renowned artist Zheng Banqiao and flocked to buy the fans from the old lady, which were sold out quickly.

Zheng painted *lan* for over 50 years. According to Xu Beihong (徐悲鸿; 1895–1953), a well-known modern Chinese painter and teacher, Zheng Banqiao is the most outstanding painter and poet in China of the last three hundred years. His paintings also made a strong impact on post-Qing calligraphers and painters, including Wu Changshuo (吴昌硕; 1844–1927), Qi Baishi (齐白石; 1863–1957), and Pan Tianshou (潘天寿; 1897–1971). Selected paintings of Zheng's works are presented in Figs. 2.18–2.23.

Zheng was one of the very few scholars good in all three arts: poetry, painting, and calligraphy. He was not only good but particularly revolutionary in his calligraphy. His special style of calligraphy made him a pioneer in the introduction of a change in Chinese calligraphic style. For example, Zheng created a unique

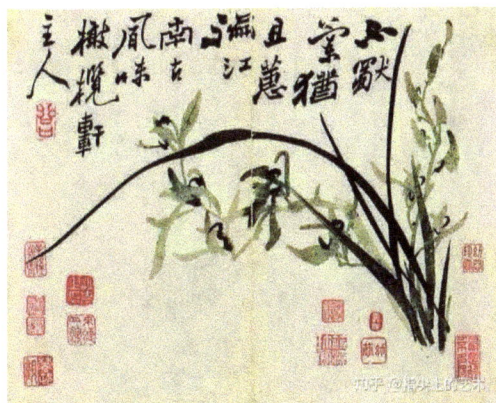

Fig. 2.18.　Painting by Zheng Banqiao (郑板桥) (1693–1765).

Fig. 2.19.　Painting by Zheng Banqiao (郑板桥) (1693–1765).

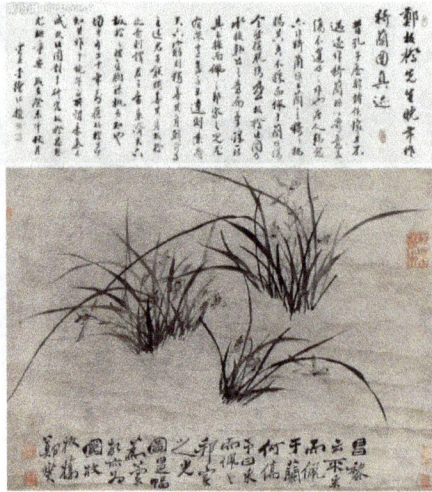

Fig. 2.20.　Painting by Zheng Banqiao (郑板桥) (1693–1765).

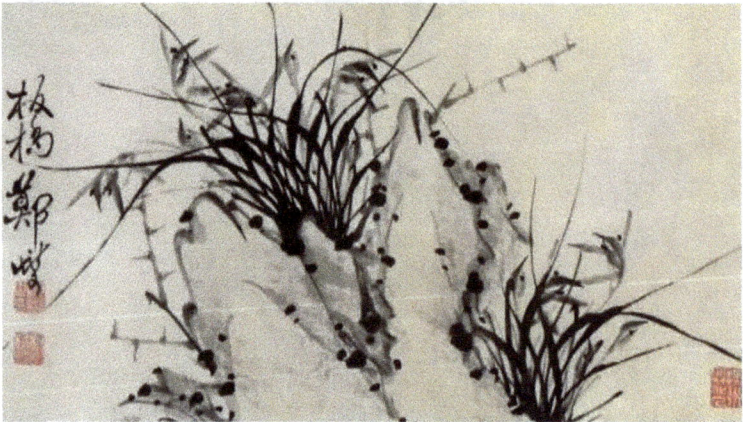

Fig. 2.21.　Painting by Zheng Banqiao (郑板桥) (1693–1765).

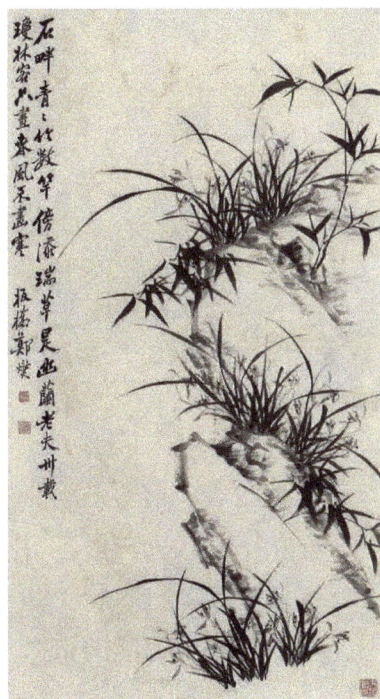

Fig. 2.22.   Painting by Zheng Banqiao (郑板桥) (1693–1765).

style of calligraphy, which he called the "*Six and a Half Calligraphy*" (六分半书). When developing his new and unique form of calligraphy, he adopted the strokes technique in the painting of Chinese orchids. The *Six and a Half Calligraphy* style is a combination of clerical script and regular script of Chinese calligraphic style.[74]

"*Nan De Hu Tu*" (难得糊涂) is one of Zheng's most famous calligraphy works[47,75,81] (Fig. 2.24). These four words, *Nan De Hu Tu*, are rich in philosophy. Different people may have different opinions and interpretations for it, with the most widely accepted being that smart people sometimes pretend to be foolish to fool others. These four words are a good reflection of Zheng's life. Till today, Chinese scholars and officials like to engrave these four words on plaques

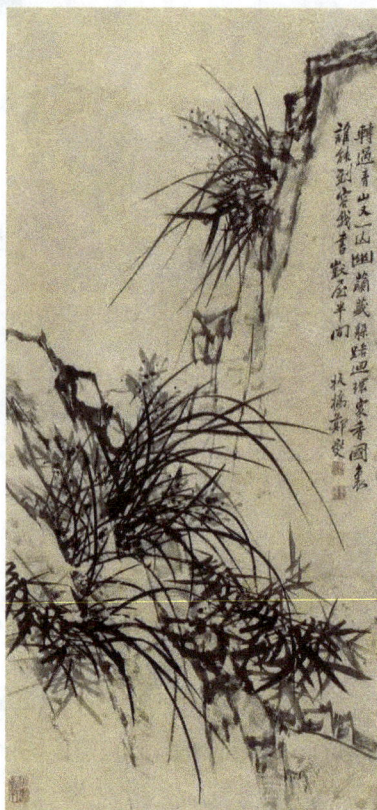

Fig. 2.23.   Painting by Zheng Banqiao (郑板桥) (1693–1765).

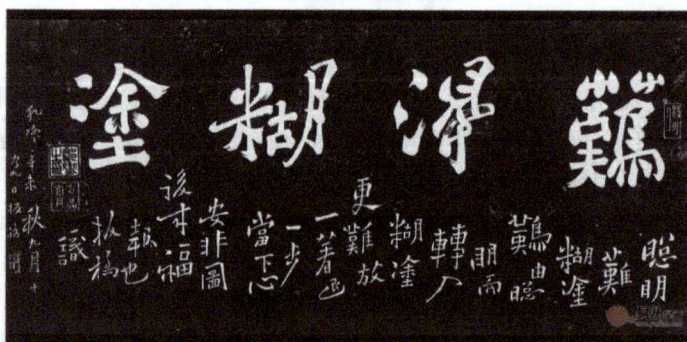

Fig. 2.24.   "*Nan De Hu Tu*" (难得糊涂), a stone rubbing of the famous calligraphy work of Zheng Banqiao (1693–1765).

and hang them on the wall at home or in the office, to remind them of the attitude and philosophy in life shared by Zheng.

A memorial hall named after him was built in Xinghua (兴化) to commemorate his outstanding works[72] (Figs. 2.25–2.26). On the 300th anniversary of the birth of Zheng Banqiao on 22 November 1993, the Ministry of Posts and Telecommunications of China also issued a set of special stamps in memory of his work on Chinese cymbidiums[73] (Fig. 2.27).

Fig. 2.25.   Zheng Banqiao Memorial Hall at Xinghua, Jiangsu.

Fig. 2.26.   A stone statue of Zheng Banqiao.

Fig. 2.27. The Ministry of Posts and Telecommunications of the People's Republic of China issued special stamps on 22 November 1993 in memory of Zheng Banqiao's work on Chinese *Cymbidium*.

Calligraphies and other artworks

Chinese calligraphy is another visual art form that *lan* is often associated with. It is a common practice for Chinese painters to incorporate calligraphy in their paintings. The subject matter of the calligraphy is often words of wisdom, poems, odes, encouraging words or selected short excerpts from literary works. It is only with a good combination of painting, calligraphy and poetry that a painting can be regarded as truly perfect.

Figure 1.1 shows the calligraphy "*A Gentleman of Noble Virtue and King of Fragrance*" (君子之风, 王者之香), which reflects praise for *lan* by Confucius. Another piece of calligraphy on *lan* is entitled "*Soul of Lan*" (兰魂) (Fig. 2.28). This ode highlighted five elements, righteousness, purity, elegance, tranquillity, and harmony, that form the *Soul of Lan*.[57]

Generally, the written inscription complements the painting. However, some painters like to vent their dissatisfaction through their calligraphy in the painting. In Zheng Banqiao's painting, for example, the inscription praised *lan* as a noble and humble gentleman and how it harmoniously coexisted with bamboos

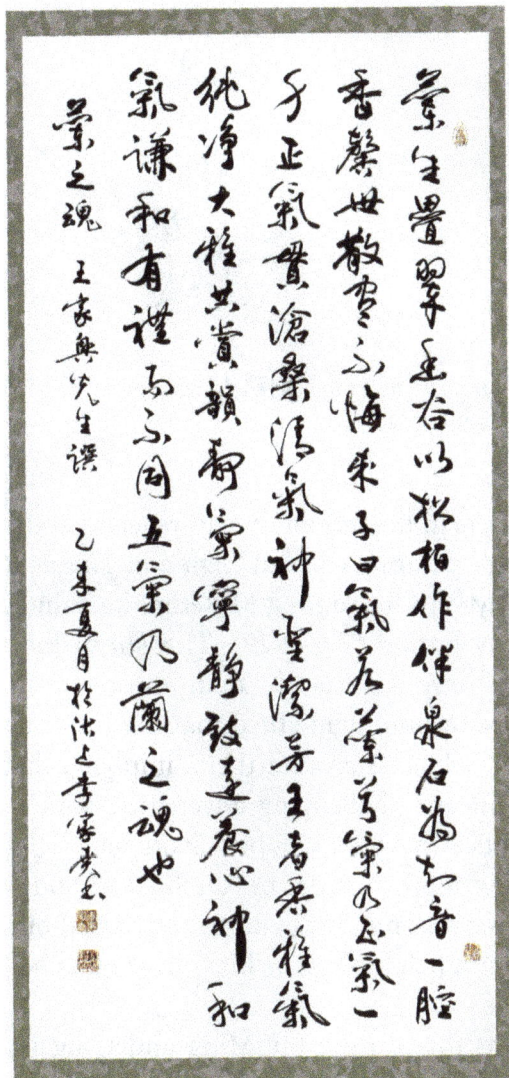

■ 王家兴/文　李家尧/书　（150×50）

Fig. 2.28. "*Soul of Lan*" (兰魂). Script: Wang Jiaxing (王家兴) (China)
Calligrapher: Li Jiayao (李家尧) (China)

Fig. 2.29.　"*Lan and Thorns Bush*" (荆棘丛兰图) painted by Zheng Banqiao is currently kept in Nanjing Museum.

amongst boulders in the mountain. However, in another horizontal scroll of Zheng's painting entitled "*Lan and Thorns Bush*" (荆棘丛兰图), currently kept in Nanjing Museum, he painted *lan* growing amongst thorny bushes (Fig. 2.29). The *lan* in the thorny weeds were growing more luxuriantly than those outside the thorny bushes. The inscription highlighted that *lan* continue to grow well even with thorny bushes around them, implying that a gentleman could withstand any challenging adversities, which drove one to achieve an outstanding and noble character.[47,74,80] On the other hand, Zheng Sixiao, the painter of "Orchids without root and soil", through the inscription in his painting, expressed his strong feeling of frustration and helplessness as his country was being ravaged by the Mongolians.

It is evident that during the Ming and Qing Dynasties, there were many works of literature, poetry, painting, and calligraphy with Chinese *Cymbidium* as the main theme. These works of art have become a valuable Chinese orchid legacy for later generations to enjoy and appreciate.

Lan is also used extensively as a motif for other objects of art, such as door stones, wood carvings, screens, fine porcelain, exquisite embroideries, foldable bamboo fans, snuff bottles, porcelain eggs and

Fig. 2.30.   Qing Dynasty snuff bottle, with painting of "*Lan*", an artwork in 1892.[82]

Fig. 2.31.   Bamboo folding fan with painting of "*Lan*".

wine pots (Figs. 2.30–2.31). In Chinese traditional architecture, the door stone is a distinctive part of the door of a building. Auspicious Chinese images like dragons or *lan* are usually carved on the door stone. Similar to how the dragon symbolises status and power, the Chinese *Cymbidium* evokes peace, harmony and happiness.

To promote public awareness of the precious orchid resources in China, the Ministry of Posts and Telecommunications of the People's Republic of China issued a set of four commemorative Chinese cymbidiums stamps on 25 December 1988. The pattern of the stamps was a selection of the valuable representative varieties, namely, *C. goeringii* "Long Zi" (龙字), *C. faberi* "Da Yi Pin" (大一品),

Fig. 2.32.   Four Chinese cymbidiums. (The stamps were issued by The Ministry of Posts and Telecommunications of the People's Republic of China on 25 December 1988). *C. goeringii* "Long Zi" (龙字), *C. faberi* "Da Yi Pin" (大一品), *C. sinense* "Yin Bian Mo Lan" (银边墨兰), *C. ensifolium* "Da Feng Wei" (大凤尾).

C. sinense "Yin Bian Mo Lan" (银边墨兰), and *C. ensifolium* "Da Feng Wei" (大凤尾)[47,53] (Fig. 2.32).

2.4. *Lan* and Traditional Chinese Customs

Customs are traditional habits or practices that people develop in certain regions or countries. The love and respect for *lan* have evolved to form an important part of Chinese traditional customs. There are many interesting anecdotes, customs, and protocols about Chinese *Cymbidium*, even in today's society.[47,57]

For example, it is a custom for people living in the southeast region of Fujian province to first build a *lan* garden to make sure the plants grow well before starting any major construction to build their new homes. This is based on the belief that if the environmental conditions are good for *lan*, it would also be good for people. Many Chinese believe in "*Feng Shui*" (geomancy). They are particular in choosing the location, direction and environment where their houses are to be built. They believe a place with good *Feng Shui* will bring them luck, fortune and health. This practice is quite widespread. In Yunnan, for instance, the ethnic minority Bai people have the custom of planting *lan* collected from the wild in the courtyard to ward off evil spirits.

After the Song Dynasty, areas around Jiangsu and Zhejiang became the economic and cultural centres of China. It was customary for the household of many wealthy families and high-ranking officials to have one pot with 6–7 stalks of *C. faberi* placed in front of Chinese calligraphy or paintings in their grand halls. *C. faberi* was their favourite orchid because the flowering stalks were tall and straight, and each stalk had as many as 12 flowers. When all flowers are in bloom, it makes a beautiful sight indeed. Chinese people also believe *lan* have spirits. If the flowers of their *lan* are growing vigorously and are in full bloom in a certain year, they think that this is a good omen, for this will bring happiness and good health to their families throughout the year. If the plant has multiple stalks in blooms or blooms off-season, it is considered to be even more auspicious.

When a person moves into a new house, relatives and friends will bring a pot of *lan* and place it in the centre of the praying table. The belief is that the fragrance of the *lan* will bring prosperity and good life. In the olden days, Chinese people also believed that the fragrance of Chinese *Cymbidium* could dispel illness and evil spirits. In the Jiang (江), Zhe (浙), and Min (闽) regions, the custom of having a pot of *lan* leading the marriage procession when marrying a daughter off is still observed. It signifies that the bride is as pure, sweet, and virtuous as the *lan*. The *lan* symbolises the parents' hope that their daughter will have a good life and be full of happiness after marriage.[47,57]

A similar custom of giving a pot of *lan* to someone moving into a new house/office or getting married is also practiced in Korea. Additionally, in Japan, when people visit patients in hospitals, they like to bring along Chinese *Cymbidium* as a gift. Not only does the flower have a sweet and delicate fragrance, but more importantly, the pollen is clamped together into masses (pollinia), and there is no dispersal of pollen that could cause an allergic reaction in the patient.[57]

A story that *lan* could induce childbirth was first published in *Jin Zhang Lan Pu* published during the Song Dynasty.[47,57] When Queen Jiading (嘉定皇后) was due to deliver her child, she experienced great pain. Traditional Chinese medical doctors summoned by the King were at a loss until a senior officer presented a pot of *C. ensifolium* to the Queen. Its sweet fragrance took the Queen by surprise, and she was overwhelmed with joy and excitement. She became less tense, and the baby was born easily. The news of the magical effect of *lan* for induction of childbirth spread far and wide. Even today, in Fujian, Sichuan, and Dali Bai (大理白族), people still follow the custom of keeping a few pots of *C. ensifolium* in the courtyard of their house when a woman is pregnant or is due for delivery. Another reason for keeping a pot of *lan* in a home is the similar pronunciation in Chinese of "*lan*" (兰) and "*nan*" (男; boy), hoping that the new-born baby is a boy. In the past, having a boy was very important as the boy would carry on the family name.[47,57]

In recent years, many schools in China have introduced programmes to get students better acquainted with the *lan culture*. In Minxi Liancheng County (闽西连城县), for example, students are encouraged to set up a *lan* garden in schools. The benefits of this programme are not only in cultivating students' interest in gardening but also increasing their awareness of the importance of conservation. Cultural activities such as collecting, growing, appreciating, painting, photographing, studying, and writing about *lan* are regularly organised for students. These activities integrate *lan culture* into the school's moral education. More importantly, students get to know better the nobility and moral integrity symbolised by *lan*.[57] There is a Chinese saying, "养兰而养于兰"[54], which means that while we are cultivating and admiring the elegance of the *Lan*, we must also learn from its noble spirit (upright, humble, and virtuous) to practise noble morality.

2.5. *Lan* as a Cultural Symbol

Over the years, *lan* has evolved from being merely a botanical term to a term that embraces and symbolises a special cultural meaning in China. It is used to describe things or objects that are elegant and beautiful. For example,

Lan dian (兰殿) — grand and elegant palace
Lan shi (兰时) — auspicious time
Lan gui (兰闺) — an elegant lady's bedroom
Lan fang (兰芳) — noble virtue
Lan zhang (兰章) — a beautiful piece of writing

Lan hun (兰魂) — lofty spirit
Lan yu (兰友) — good and trustworthy friend
Lan ke (兰客) — distinguished guests
Lan yan (兰言) — intimate conversation
Lan yin (兰音) — beautiful voice
Lan jie (兰姐) — sworn sister

Lan is also popularly used in naming Chinese girls, such as "Mei Lan" (美兰; beautiful *lan*), "Su Lan" (素兰; pure *lan*), or "Lan Xiang" (兰香; fragrant *lan*). By associating the names of their daughters with "*Lan*", parents wish for their daughters to have a heart as pure as *lan*.[47,64] *Lan* has become a symbol of beauty, happiness, and auspiciousness.

In Chinese culture, there are other similar examples. The word "*Long*" (龙; Chinese dragon), for example, has also transcended its original meaning and become a cultural symbol. *Long* is a legendary creature in Chinese mythology, folklore, and culture. In the Orient, dragons symbolise wisdom, vitality, power, strength, and prosperity. In the olden days, the Emperor of China usually used the dragon as a symbol of his imperial strength and power. For "*long*", we have "*long che*" (龙车; imperial chariot), "*long pao*" (龙袍; emperor's court dress), "*long ju*" (龙驹; fine horses/brilliant young man), and others.[64]

Chapter 3

Five Common Chinese *Cymbidium* Species

Chinese cymbidiums are members of the genus *Cymbidium* Sw. which are widely distributed in China, India, Japan, Korea, Australia, and Malaysia. The genus *Cymbidium* Sw. is divided into three subgenera, namely, *Cymbidium, Cyperorchis,* and *Jensoa.* Chinese cymbidiums, a group of terrestrial orchids, belong to the subgenus *Jensoa.*[7,65] In China, Chinese *Cymbidium* is mainly found in provinces such as Jiangsu, Zhejiang, Fujian, Sichuan, Yunnan, Shanxi, Hainan, Anhui, Guangdong, Guangxi, Henan, and Taiwan.[65]

The growth conditions and cultivation of different types of Chinese cymbidiums varied depending on different geographical distributions, climates, and ecological environments. The five most well-known Chinese *Cymbidium* species are *Cymbidium goeringii* (春兰), *Cymbidium sinense* (墨兰), *Cymbidium ensifolium* (建兰), *Cymbidium faberi* (蕙兰), and *Cymbidium kanran* (寒兰).

Over the years, through natural selection and hybridisation programmes, many new cultivars/varieties have been produced in each of these five Chinese cymbidiums with considerable variations in floral and leaf forms. Different cultivars/varieties are grouped into various categories/types based primarily on their flower forms or shapes and leaf variations. A brief description of the five common Chinese *Cymbidium* species and varieties is given below[60,61,65]:

3.1. *Cymbidium goeringii* (春兰)

C. goeringii, also known as "*Spring Orchid*", has a long cultural and cultivation history in China. It is one of the earliest popular Chinese *Cymbidium* species. The appreciation and judging criteria of Chinese cymbidiums, in many respects, were based originally on the features and characteristics of *C. goeringii*. In China, this species is commonly found at an altitude of 300–2200 m above sea level, and it grows in an array of habitats from rocky slopes to open spaces in forests.

 C. goeringii has an inconspicuous ovoid shape and flattened pseudobulbs. It has 4–7 grass-like leaves, each being 20–40 cm long and about 6–9 mm wide. Like other Chinese cymbidiums, it has many long thick, fleshy and white roots. The flower stalk is short and graced with a single flower, and rarely with two. That unique characteristic of *C. goeringii* makes it easy to identify. Its flowering period is from January to March, and the flowers have an exquisite fragrance. The colours of the flowers are mainly green or dark green, but can also be yellow, red, white, brown or green with purplish venation. *C. goeringii* has about 200 varieties and is divided into two major groups — *Chinese Spring Orchids* and *Japanese Spring Orchids*. The former stands out with its flower shape and fragrance, whereas the latter is known for its flower colour.

 C. goeringii is one of the most widely cultivated orchids in China. The main reasons for its popularity are as follows:

1. It is easy to grow.
   C. goeringii has rather simple growth requirements, making it suitable for novice growers. It thrives well, if it is in a humid and warm environment and away from direct sunlight and freezing temperatures in winter.
2. It has an elegant plant form.
   The *C. goeringii* is relatively small and occupies little space. The shape of the leaves is half arched, green, and shiny, making it a perfect match with traditional Chinese furniture. Pots of *C. goeringii* can be seen on the patios of many houses along the

Yangtze River Basin in China. When grown in specially designed flowerpots, it fits very well into the decor of the study or living room of a traditional Chinese home.

3. It is a good subject for Chinese paintings.

Its green, slightly curved, or drooping shiny leaves are the basic standard form for painting Chinese *Cymbidium* leaves in China.

The well-known varieties include *C. goeringii* 'Song Mei' (宋梅), *C. goeringii* 'Wang Zi' (汪字), *C. goeringii* 'He Shen Mei' (贺神梅), *C. goeringii* 'Xue Shan' (雪山), *C. goeringii* 'Lu Yun' (绿云), *C. goeringii* 'Xi Shen Mei' (西神梅), and *C. goeringii* 'Long Zi' (龙字) (Figs. 3.1–3.8).

Fig. 3.1. *C. goeringii* 'Song Mei' (宋梅).

Fig. 3.2. *C. goeringii* 'Long Zi' (龙字).

Fig. 3.3.   *C. goeringii* 'Xi Shen Mei' (西神梅).[60]

Fig. 3.4.   *C. goeringii* 'Wang Zi' (汪字).

Fig. 3.5.   *C. goeringii* 'Xue Shan' (雪山).

Fig. 3.6.   *C. goeringii* 'Da Fu Gui' (大富贵).

Fig. 3.7.   *C. goeringii* 'Jin Tai He' (金泰荷).

Fig. 3.8.   *C. goeringii* 'Yu Guan He Ding' (宇冠荷鼎).

3.2. *Cymbidium sinense* (墨兰)

The *C. sinense* is another popular Chinese ornamental orchid with a long history of cultivation in China. It has been recorded in *Jin Zhang Lan Pu* as early as the Southern Song Dynasty.[60] As this species has many flowers per stalk blooming around the Chinese New Year, it is called "*Bao Sui Lan*" (报岁兰/ 拜岁兰), meaning "Heralding New Year Orchid".

    C. sinense is mainly found in humid lowland forests with well-drained rich soil, and on mountain slopes at an altitude below 2000 m. It prefers warmer temperatures of around 18–22°C. The pseudobulbs are either elliptical or ovoid, ranging from 1.5 to 2.5 cm in diameter, but generally larger than that of the *C. ensifolium*. It normally has 3–5 dark green leathery leaves, 40–70 cm in length. The flowering period is from October to March of the following year. The inflorescence is robust and fragrant and has many flowers (from 7 to as many as 20 per stalk). Their colour varies from dark purple to nearly black with light-coloured lips, and it is for this very reason that *C. sinense* is commonly known in China as "*The Ink Orchid*" (墨兰). During the Song Dynasty, dark purple was the preferred colour in judgement and appreciation of Chinese cymbidiums. In the 1950s, many high-quality varieties with different leaf and flower forms that originated in Hong Kong and Taiwan were sold in Chinese *Cymbidium* markets in Guangdong. This had a major impact on the cultivation and appreciation of *C. sinense* in China.

    C. sinense flowers are generally not gorgeous and charming, but their unique shiny dark green leaves and clear veins give people a refreshing and lively feeling. Today, many of its varieties are specially selected for their exotic and beautiful leaf shape and variegation. This allow orchid lovers to enjoy the beauty and charm of *C. sinense* even when the plant is not in bloom. Having beautiful leaves makes this orchid very popular particularly in Guangdong and Southeast Asia.

Some of the famous varieties are *C. sinense* 'Hui Mo' (徽墨), *C. sinense* 'Yun Nan Bai Mo' (云南白墨), *C. sinense* 'Da Mo' (达摩), *C. sinense* 'Nan Hai Su Qi' (南海素奇), *C. sinense* 'Qui Bang' (秋榜), and *C. sinense* 'Jin Bian Mo' (金边墨) (Figs. 3.9–3.18).

Fig. 3.9.   *C. Sinense* 'Qui Bang' (秋榜).[60]

Fig. 3.10.   *C. sinense* 'Yun Nan Bai Mo' (云南白墨).[60]

Fig. 3.11.   *C. sinense* 'Qi Hei' (墨兰企黑).

Fig. 3.12.   *C. sinense* 'Nan Hai Su Qi' (南海素奇).

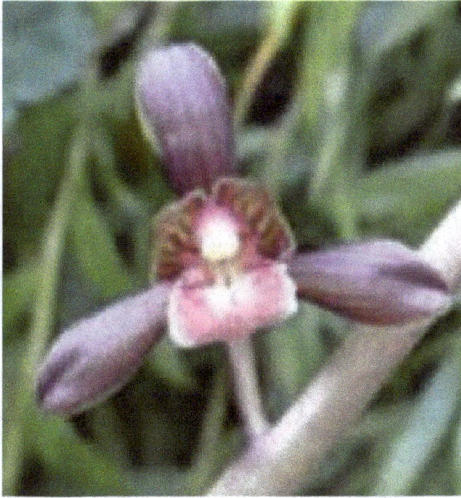

Fig. 3.13.   *C. sinense* 'Yuan He' (圆荷).

Fig. 3.14.   *C. sinense* 'Tao Ji' (桃姬).

Fig. 3.15.   *C. sinense* 'Dong Fang Hong' (东方红).

Fig. 3.16.   *C. sinense* 'Qiu Xiang' (秋香).

Fig. 3.17.   *C. sinense* 'Hong Tai Yang' (红太阳).

Fig. 3.18.   *C. sinense* 'Qiu Bang Bai Mo' (秋榜白墨).

3.3. *Cymbidium ensifolium* (建兰)

C. ensifolium is known as "*Four-Seasons Lan*" (四季兰) because it blooms freely throughout the four seasons of the year. However, its main flowering period is from June to October. It is often found at an elevation 600–1800 m above sea level on the slopes of temperate forests with well-drained soils and is widely distributed throughout Indochina, China, Borneo, Papua New Guinea, the Philippines, and Japan.

C. ensifolium is also commonly known as "*Jian Lan*" (建兰). Fujian province was the major place of cultivation of *C. ensifolium* in the past and has remained so today. It is believed that the word "*jian*" (建) in "*jian lan*" (建兰) is derived from the word "Fujian" (福建).

Its pseudobulbs are elliptic and inconspicuous with 2–6 light leathery leaves (length: 30–60 cm; width: 1–2 cm). It usually produces 5–9 flowers per inflorescence, but the number can go up to 13. However, the flower stalks are 25–40 cm, with well-spaced flowers on the raceme. While variable in colour, the *C. ensifolium* usually has a light yellowish-green shade and with violet-red stripes and spots. This species is known for its exquisite scent, inflorescence habit, and flower count as well as its compact growth form.

Amongst all the Chinese cymbidiums, *C. ensifolium* is relatively simple to grow and maintain. It grows fast, flowers freely, has a long flowering period, and a high adaptability to the environment. The price is generally affordable to the public. For those who wish to start growing Chinese cymbidiums, *C. ensifolium* has often been recommended as the first choice. It is an ideal plant for growing on the balcony, living room, or courtyard. For example, *C. ensifolium* 'Xiao Tao Hong' (小桃红), *C. ensifolium* 'Hong Yi Pin' (红一品), and *C. ensifolium* '*Long Yan Su*' (龙岩素) are very popular amongst novice orchid growers.

From the time of the Song Dynasty to the Qing Dynasty, *C. ensifolium* with a pure colour flower was well sought after. For example, *C. ensifolium* 'Yong Fu Su' (永福素) produced in Yong Fu Town, Zhang Ping (漳平), was once awarded as a treasure flower by the Jiajing Emperor (嘉靖皇帝) in the Qing Dynasty.[61]

C. ensifolium, C. goeringii, and C. faberi have been reported to have medicinal value in *Ben Cao Gang Mu,* the most celebrated Chinese herbal text published during the Ming Dynasty. *C. goeringii* is used to treat neurasthenia, roundworm infection and haemorrhoids; *C. ensifolium* is good for treating patients with weakness in the lung; while *C. faberi* is used for the treatment of gynaecological diseases.[57]

Many of the *C. ensifolium* varieties are found in the Fujian, Guizhou, Zhejiang, Yunnan, Guangxi, Hainan, Taiwan, and Guangdong regions. After decades of cultivation and breeding, numerous new varieties have been produced in China. These include *C. ensifolium* 'Xia Huang Mei' (夏皇梅), *C. ensifolium* 'Tie Gu Su Xin' (铁骨素心), *C. ensifolium* 'Huang Guang Deng Mei' (黄光登梅), *C. ensifolium* 'Zi Yun' (紫云), *C. ensifolium* 'Hong Yi Pin' (红一品), and *C. ensifolium* 'Yi Pin Mei' (一品梅) (Figs. 3.19–3.28).

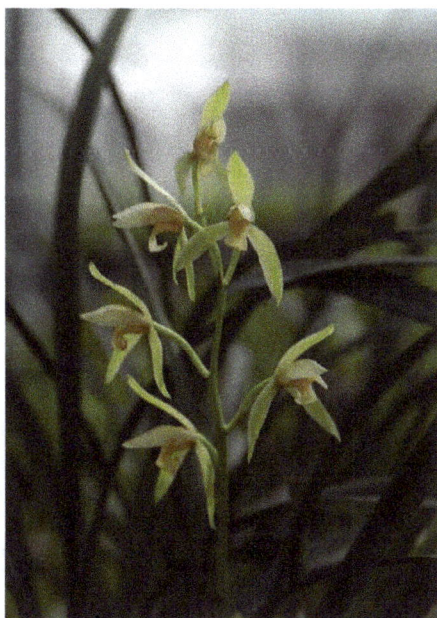

Fig. 3.19.   *C. ensifolium* 'Tie Gu Su Xin' (铁骨素心).

Fig. 3.20.   *C. ensifolium* 'Huang Guang Deng Mei' (黄光登梅).

Fig. 3.21.   *C. ensifolium* 'Zi Yun' (紫云).

Fig. 3.22.　*C. ensifolium* 'Xiao Guo Hun' (小国魂).

Fig. 3.23.　*C. ensifolium* 'Hong Niang' (红娘).

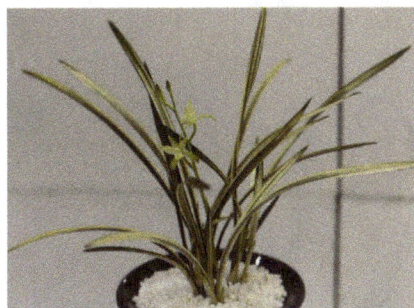

Fig. 3.24.　*C. ensifolium* 'A Li Shan Xue Yu' (阿里山雪玉).

Fig. 3.25.   *C. ensifolium* 'Chi Cheng' (赤诚).

Fig. 3.26.   *C. ensifolium* 'Zhong Hua Cai Xia' (中华彩霞).

Fig. 3.27. *C. ensifolium* 'Ming Shan Yan Zhi' (名山胭脂).

Fig. 3.28. *C. ensifolium* 'Ling Long He' (玲珑荷).

3.4. *Cymbidium faberi* (蕙兰)

C. faberi and *C. goeringii* share similar habitats in nature and therefore, their horticultural requirements are rather similar. *C. faberi* is relatively cold tolerant and grows well in damp but well-drained mountain slopes and open shrubby places at 1000–3300 m above sea level.

C. faberi generally has nine flowers per inflorescence and is known as the "*Nine Children Lan*" (九子兰). Some may have as many as twelve flowers per stalk. Its inflorescence is sub-erect and slightly curved. *C. faberi* blooms from March to May. Its scent is less fragrant than that of *C. goeringii*. The pseudobulb is inconspicuous and has 5–7 (~9) slender leaves (length, 25–75 cm; width, 7–12 cm) with transparent veins (particularly the mid-vein). The leaf margins are sharply serrated, and roots are long and thick. Its petals, as well as sepals, are light yellow-green with purplish-red patches on the lip.

C. faberi was well documented in the book "*Di Yi Xiang Bi Ji*" published in 1796 during the Qing Dynasty with a good description of its cultivation methods and care.[47] After the Song Dynasty, the Jiangzhe area (江浙) became the centre of China's economic and cultural development. Many wealthy families and high-ranking government officers had households with spacious halls and large gardens, and it was a tradition to have a pot of *C. faberi* placed near the plaque or couplets in the main hall of the houses. *C. faberi* has an elegant and attractive plant form when all flowers are in bloom. It is common to have 5–7 spikes of *C. faberi* in one pot blooming at the same time. When all the flowers are in full bloom, it makes an eye-catching display at a large hall, courtyard, teahouse, and restaurant (Figs. 3.29–3.31). Today, most people in Jiangsu and Zhejiang continue to maintain the tradition of growing a few pots of *C. faberi* in their residences.

In recent years, many novel varieties of this orchid with attractive flower forms, shapes, and colours have appeared on the market.

Some of the popular varieties include *C. faberi* 'Cheng Mei' (程梅), *C. faberi* 'Yuan Zi' (元字), *C. faberi* 'Cui Mei' (崔梅), *C. faberi* 'Lao Shang Hai Mei' (老上海梅), *C. faberi* 'Jiang Nan Xin Ji Pin' (江南新极品), and *C. faberi* 'Da Yi Pin' (大一品) (Figs. 3.29–3.40).

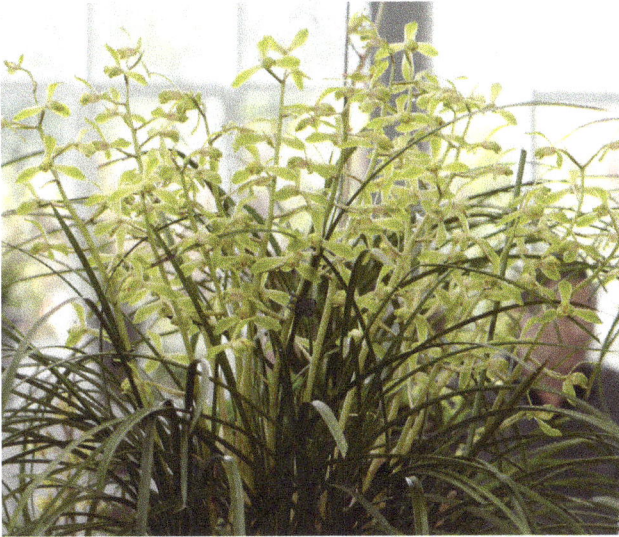

Fig. 3.29.   *C. faberi* 'Jie Pei Mei' (解配梅).

Fig. 3.30.   *C. faberi* 'Cui Mei' (崔梅).[60]

Fig. 3.31.   *C. faberi* 'Yuan Zi' (元字).

Fig. 3.32.   *C. faberi* 'Yuan Zi" (元字).

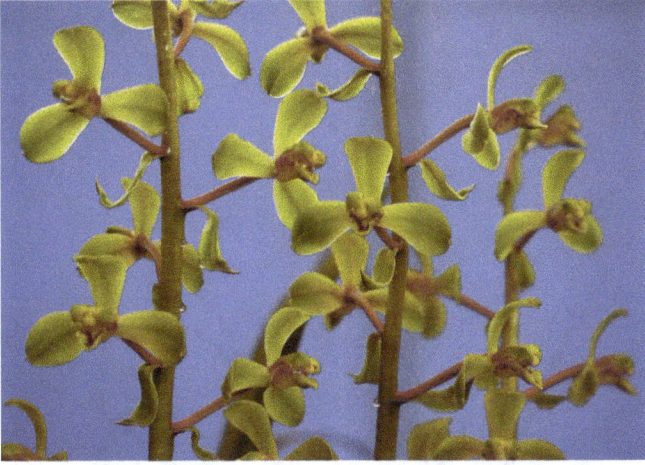

Fig. 3.33.　*C. faberi* 'Cheng Mei' (程梅).

Fig. 3.34.　*C. faberi* 'Guan Ding' (关顶).

Fig. 3.35.   *C. faberi* 'Jiang Nan Xin Ji Pin' (江南新极品).

Fig. 3.36.   *C. faberi* 'Lao Ji Pin' (老极品).

Fig. 3.37.   *C. faberi* 'Da Yi Pin' (大一品).

Fig. 3.38.   *C. faberi* 'Tao Bao' (陶宝).

Fig. 3.39.   *C. faberi* 'Hua Xia Xin Mei' (华厦新梅).

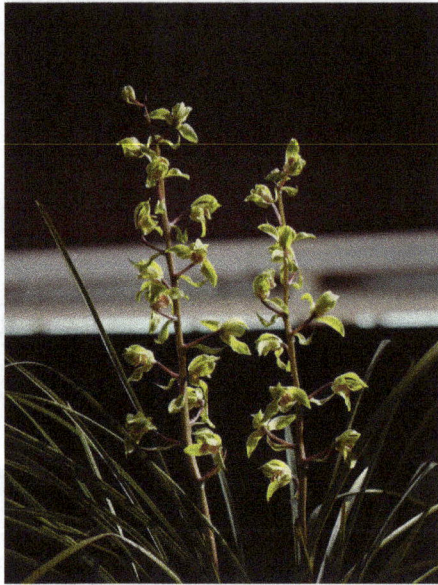

Fig. 3.40.   *C. faberi* 'Er Wan Yuan' (二万圆).

3.5. *Cymbidium kanran* (寒兰)

C. kanran is known as "*Han Lan*" (寒兰; *Cold Lan*) in Chinese. The plant is native to the southern parts of China, Korea, and Japan. It is found on damp rocky soil in evergreen broad-leaved mountain forests at an altitude of 400–2400 m above sea level. The evergreen broad-leaved forest not only provides a shield for the plants from the scorching sun during summer but also blocks the cold wind and snow in winter.

The plant is medium sized and has narrow ovoid conspicuous pseudobulbs with 3–7 long, slender, erect leaves (length: 35–70 cm; width: 1–2 cm), and its leaf tips are weakly serrated. The inflorescence is 25–70 cm long and grows above the leaves. *C. kanran* has as many as 10 flowers on one stalk. The flowering season is from October to January of the following year. Orchid lovers are attracted to the overall beauty, slenderness and symmetry of this plant. The reason why it is so much admired by people is because it has a well-proportioned and elegant leaf posture, gorgeous flower colour, mellow and long-lasting fragrance, as well as an air of mystery and vitality.

Compared to *C. sinense* and *C. ensifolium*, the flower of *C. kanran* is very much slender but considerably richer in colour ranging from yellow, green, violet, red, and dark violet to pearl white. The flying sepal-type flowers are well spaced and come in different shapes, with some looking like dragonflies or cranes resting on the branches. They look truly attractive[57] (Figs. 3.41–3.43).

The varieties of *C. kanran* are broadly divided into two groups[57]: The first group is based on the flowering season:

1. *Spring Han Lan* (春寒兰) — flowering period from January to March; relatively few varieties
2. *Summer Han Lan* (夏寒兰) — flowering period from June to August; pleasantly scented; a limited number of varieties
3. *Autumn Han Lan* (秋寒兰) — flowering period from September to November; strongly scented; a large number of varieties
4. *Winter Han Lan* (冬寒兰) — flowering period from November to January of the following year; strongly scented; large number of varieties

The second group is based on flower colour:

1. *Green Han Lan* (青寒兰) — green sepals and petals; some with purple stripes; most common colour in Han Lan

2. *Red Han Lan* (红寒兰) — dark pink sepals and petals; lip with violet-red spots
3. *Violet Han Lan* (紫寒兰) — violet-red sepals and petals; lip with violet-red spots
4. *Mixed-Colour Han Lan* (复色寒兰) — greenish-purple sepals and petals; lip with dark purple veins and spots

Generally, the green and red Han Lan are the more precious varieties. The preferred flower colour of a good variety of *C. kanran* is single colour without variegation. One of the most famous varieties is "*Su Xin Han Lan*" (素心寒兰). It blooms in winter and exudes a pure, fresh and pleasant fragrance.

Some interesting varieties are *C. kanran* 'Qing Han Lan' (青寒兰), *C. kanran* 'Zi Han Lan' (紫寒兰), *C. kanran* 'Yang Gui Fei' (杨贵妃), *C. kanran* 'Tie Lian' (铁脸), *C. kanran* 'Bi Yi Shuang Fei' (比翼双飞), and *C. kanran* 'Qi Hei' (企黑) (Figs. 3.41–3.50).

C. kanran was first described in 1902 by Makino in Japan. It is popular in Japan and South Korea, and the people love *C. kanran* more than any other Chinese cymbidiums. It is apparent that appreciation and cultivation of *C. kanran* have a much longer history in Japan than in China. Two factors account for why it was less popular in China in the 1900s[57,65]:

1. Influence of the historical criteria used to judge the beauty of Chinese *Cymbidium* flowers in China. The same criteria have remained largely unchanged since the Qing Dynasty. They are based primarily on the rounded floral form as the norm for judging Chinese *Cymbidium* species, and the floral form of *C. kanran* is slender and is different from that of many other Chinese cymbidiums, such as *C. goeringii, C. sinsense, and C. ensifolium*.
2. Very few people like to grow *C. kanran* because it is more difficult to maintain. It has a low seed germination rate compared to other Chinese cymbidiums. It is also difficult to cultivate *C. kanran* collected from the wild in nurseries.

Fig. 3.41. *C. kanran* 'Yan Nan Fei' (雁南飞).

Fig. 3.42. *C. kanran* 'Bi Yi Shuang Fei' (比翼双飞).

Fig. 3.43.   *C. kanran* 'Qing Ting Fei Zhan' (蜻蜓飞展).

Fig. 3.44.   *C. kanran* 'Jin Qian Bao' (金钱豹).

Fig. 3.45.   *C. kanran* 'Yang Gui Fei' (杨贵妃).

Fig. 3.46.   *C. kanran* 'Tie Lian' (铁脸).

Fig. 3.47.   *C. kanran* 'Jing Gang Cai Xia' (井冈彩霞).

Fig. 3.48.   *C. kanran* 'Qiu Yue' (秋月).

Fig. 3.49.   *C. kanran* 'Xi Yang Hong' (夕阳红).

In recent years, there has been an increase in the popularity of *C. kanran* in China. This is largely attributed to a better understanding of its growth requirements, changes in criteria to judge and appreciate the unique beauty of *C. kanran*, and the availability of its new and better varieties.

3.6. The Naming of Chinese Cymbidiums

Traditionally, the varieties/ cultivars of Chinese *Cymbidium* are named after the characteristics of the flower (form or colour), breeder, place of origin, an event, fragrance, leaf art, or the name of insects, birds and animals. Interesting anecdotes in the naming of some Chinese *Cymbidium* varieties are given as follows[47,51,57,60]:

For instance, *C. ensifolium* 'Zi Gong Su' (自贡素) was named after the place of its origin. This fine-leaved Chinese *Cymbidium* was first found at the foot of Zigong Mountain (自贡山) in Sichuan in 2003.

During the Shunzhi Period (顺治年间) of the Qing Dynasty (1644–1660), a distinctive and interesting *C. goeringii* was discovered in Suzhou. The flower was emerald green (翠) in colour with a big round lip and had three plum-shaped (梅) sepals that were round like coins (钱). It was therefore named 'Cui Qian Mei' (翠钱梅). This flower was beautiful and had a delicate floral fragrance. It has been reported that *C. goeringii* "Cui Qian Mei" is no longer found these days.

'Qiu Bang' (秋榜) is a variant of *C. sinense* originated from the Pu'er area (普洱) of Yunnan Province. It is one of the top ten famous varieties of *C. sinense*. It blooms in the 9th month of the Lunar Calendar. In ancient times, the imperial court usually released the name list of candidates (榜) who had successfully passed the imperial examinations in autumn (秋). This flower got this name as it bloomed just when the imperial court released the good news of the selection of candidates. This plant was often displayed in the imperial examination hall as candidates believed this Chinese *Cymbidium* variety would bring them luck and an auspicious future.

C. faberi 'Feng Qiao' (蜂巧) is a red-shelled green variety of *C. faberi* with plum-shaped sepals flower, and has been praised as a treasure by the orchid community since the Qing Dynasty. This flower was named by the Qing Emperor Kangxi (康熙; 1654–1722). According to legend, in the fifth year of Emperor Kangxi of the Qing Dynasty in 1666, there was a theft case of a Chinese *Cymbidium* that eventually led to a lawsuit and was finally referred to the Imperial Court for judgement. Kangxi was curious about the plant that was stolen. While he was admiring the orchid, a bee happened to fly around the flower, and Kangxi casually named this orchid "Feng

Qiao" (蜂巧). Bee in Chinese is "蜂", and "巧" means "coincidence". Since then, this story has been commonly shared amongst Chinese households.

C. ensifolium 'Huang Guang Deng Mei' (黄光登梅), a yellowish plum-shaped sepal orchid, was named by a *lan* lover, Li Guangdeng (李光登) (Fig. 3.20). This Chinese *Cymbidium* grew wild at Shi Niu Lan (石牛栏), a mountainous area in Zigong Rong County (自贡荣县), Sichuan Province. A peasant woman named Madam Wu accidentally collected the orchid from the wild in 1992 and planted it beside her tattered pig trough. The beauty and fragrance of this orchid attracted the attention of Li, who also lived in the same village. He exchanged two of his Chinese cymbidiums for Madam Wu's orchid. After his tender care, the Chinese *Cymbidium* flowered profusely in the following year and became an instant attraction in the village. In 2006, it was rated as one of the eight precious varieties of *C. ensifolium.*

The thin-leaf *C. kanran* 'Shen Long Tu Zhu' (神龙吐珠) was named after the special form of its lip. Both pedicel and flower are green in colour; the calyx is emerald green. The lip is white, elongated and the front end is rounded. The whole flower is shaped like a green dragon spitting a pearl from its tongue which is indeed fascinating and is a well-sought-after variety of *C. kanran* (Fig. 3.50).

There is an interesting Chinese *Cymbidium* that originated from Guizhou with the Chinese name 'Wu Sha' (乌纱). The two lateral dark brown sepals are in a 180-degree straight line and curled up at both ends. The flower is shaped just like the hat of the ancient court officials in the Eastern Jin Dynasty (Fig. 3.51). The hat was called "wu sha mao" (乌纱帽; black gauze hat) and was made from black cotton yarn. In the Sui and Tang Dynasties, it became common amongst ordinary people to wear this hat, but the hat was completely abolished during the Qing Dynasty. However, people today still

Fig. 3.50.   *C. kanran* 'Shen Long Tu Zhu' (神龙吐珠).

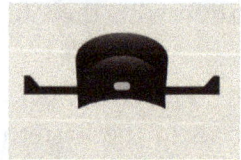

Fig. 3.51.   *C. goeringii* 'Wu Sha' (乌纱).[60]

associate this kind of hat with government officials. Those who work as officials are often referred to as "wearing a wu sha mao", and those who are dismissed by the government/company are described as "taking off the wu sha mao".

Chapter 4

Biology of Chinese *Cymbidium*

So, what is it that makes Chinese cymbidiums so appealing and special to both Chinese artists and the public? What are the features people look for when judging and appreciating Chinese cymbidiums?

Each species of Chinese cymbidiums has its characteristic features, which are the result of genetic and natural variations. The art of appreciation of Chinese cymbidiums can be summarised in four words: "*Scent, Colour, Form, and Charm*" (香, 色, 姿, 韵).[60] Scent, colour, and form (or posture) are rather easy to define and will be discussed in Section 4.1, "Structure and Characteristics of Plant Parts". The definition of charm of the Chinese *Cymbidium*, however, is quite abstract. It can be defined as the harmonious unity of external and internal beauty. It could also represent the expression of associations and feelings that people generate in their minds while observing the Chinese *Cymbidium*.

Chinese cymbidiums have unique leaf and flower structures and floral fragrances. The differences between the Chinese cymbidiums and other plants such as plum blossoms, bamboos, and pines, in the eyes of the Chinese people, were best described in the ancient book *Wang Shi Lan Pu* (王氏兰谱) written by Wang Gui Xue (王贵学) in the Song Dynasty[19,64]:

竹有节而啬花, 梅有花而啬叶,
松有叶而啬香, 惟兰独并有之。

Hu (1971)[19] translated this as follows:

"Bamboo has integrity (chieh; 节 = joints, the same character as integrity) but is short of flowers. Mei (梅; Prunus mume) has flowers but is short of leaves during flowering time. Pine has leaves when in flower but is short of fragrance. Orchid (Cymbidium) has leaves, flowers, and fragrance all at the same time."

A man of great virtue in China is commonly referred to as a *"Gentleman"*. The plum blossom, Chinese *Cymbidium*, bamboo and chrysanthemum are commonly known as the *"Four Gentlemen"* (四君子) or the *"Four Noble Plants"* in Chinese culture. That is because the natural characteristics of these four plants share certain characteristics in common with human virtues, with each of them symbolising certain esteemed morals.[47,64,70]

Plum blossoms usually bloom in winter or early spring. Hence, its tolerance of the cold is associated with a man of strong character and self-control. This is seen as the ability to thrive in an adverse environment, a feature that is admired by many ancient Chinese scholars.

Bamboos grow straight and remain green even in winter, representing a firm and tenacious character. Moreover, bamboos are hollow inside and hence are regarded as modest, honest, and upright. These characteristics have been well respected by Chinese people since ancient times.

Chrysanthemums on the other hand, are famous for their adaptability to the natural environment. Chrysanthemums can withstand the frost in late autumn and bloom with graceful shapes and bright colours. It symbolizes leisure and auspiciousness.

Chinese cymbidiums are smaller in size than the other three noble plants, but most varieties are pleasantly scented. The fragrance is never overpowering, thus symbolising modesty and gracefulness and a balanced personality. Chinese *Cymbidium* is upright, elegant, and without branching. Unlike chrysanthemums, it does not need bamboo strips to prop up its branches, and the flowers have a relatively long life. Given its elegant plant form, distinct leaf and flower structure, and pure and sweet fragrance, it is not surprising the Chinese people hold it so close to their hearts.

The *Four Noble Plants* also represent the four seasons — the Chinese *Cymbidium* for spring, bamboo for summer, chrysanthemum for autumn, and plum blossom for winter. During the Song Dynasty, the *Four Gentlemen* were popularly featured in Chinese paintings. They also appear as a group in four foldable screens placed in the main halls of Chinese homes, restaurants, or other public places.

4.1. Structure and Characteristics of the Plant Parts

Chinese cymbidiums belong to a highly evolved group of plants, and the plant parts have undergone structural adaptations. These include the root, stem, leaf, and flower. We will examine the structural adaptions in Chinese cymbidiums and discuss how these affect physiological functions, such as water conservation, drought resistance, mineral uptake, and carbon metabolism.

Pseudobulbs

The pseudobulb is a unique feature in many orchids. It is derived from the thickening of parts of the shoot between the nodes. Each pseudobulb may be composed of one or more internodes.[2] It is a swollen fleshy body shaped like a bulb from which all the leaves and flower stalks grow (Fig. 4.1). Like many of the orchids, Chinese cymbidiums also have pseudobulbs, but they are generally shorter, less conspicuous, and tend to cluster at the base of plants.

Pseudobulbs are modified stems with thick cuticles. Unlike the stems, there are no stomata present in the orchid pseudobulb.[14,88]

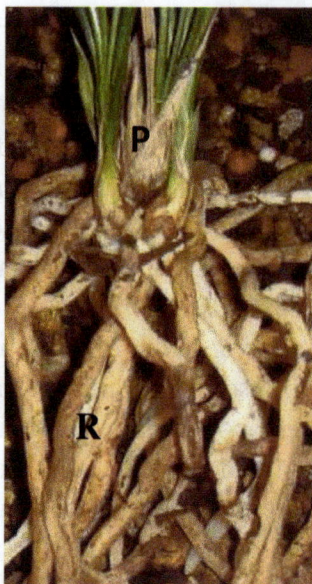

Fig. 4.1.   Pseudobulbs and roots of Chinese *Cymbidium* (**R**, root; **P**, pseudobulb).

Intact *Oncidium* orchid pseudobulbs show no carbon dioxide gas exchanges in light and in darkness. However, carbon dioxide evolution can be detected in the dark after partial removal of the cuticle (2 cm × 2 cm).[14] Evidently, the development of the water conservation feature in orchid pseudobulbs with an impermeable layer of cuticle and the absence of stomata is done at the expense of CO_2 diffusion. Orchid pseudobulbs serving as water and food storage organs in orchids have been extensively studied.[13,14]

Roots

Although they are terrestrials, Chinese cymbidiums have long, fleshy, and thick roots, like other epiphytic orchids[7,65] (Fig. 4.1). They have a central core consisting of cortex and stele and are covered with white spongy dead tissue known as velamen (Fig. 4.2). A highly specialised layer of cells called the exodermis lies between the cortex

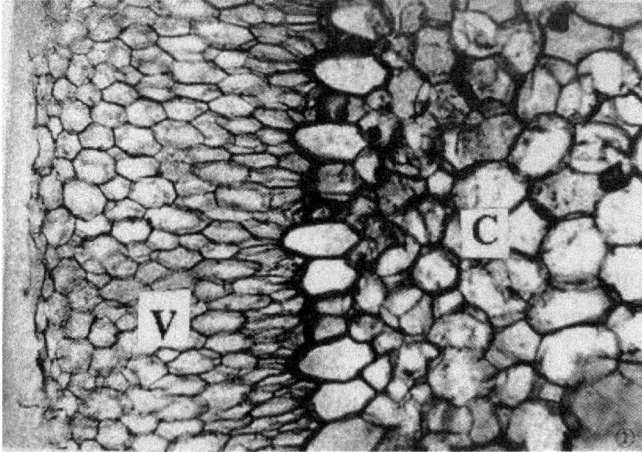

Fig. 4.2. Scanning electron micrograph of a cross section of *C. sinense* root (**V**, velamen; **C**, cortex).

and the velamen. The velamen serves as a sponge to absorb and hold water and nutrients, which are released slowly into the central core via passage cells present in the exodermis. The presence of velamen in orchid roots is an adaptive feature for drought resistance.[14,88]

Having creamy-white healthy roots is therefore the pursuit of all Chinese *Cymbidium* growers. Poor root growth affects leaf growth and flower development of Chinese cymbidiums, and it is attributed mainly to pathogens attacking the root system. Selecting a suitable potting mix is one important way of promoting a healthy root system. With better water drainage and air movement in the potting mix, the roots can grow better and healthier, thus reducing the chance of infection by pathogens.

In recent years, research on tropical orchids showed that the orchid root is not only an organ for the absorption of minerals but is also an active site for the biosynthesis of important plant growth hormones. Such hormones are involved in promoting flower development and for drought resistance.[14,42] These studies have further demonstrated the importance of orchid roots to overall plant growth.

Leaves

Chinese cymbidiums have relatively long slender leaves with pointed leaf tips. The width of the leaves varies from species to species. For example, the leaf width of *C. sinense* is 2–3.5 cm, while that of *C. kanran* is 1–1.5 cm.[60] The leaf texture and appearance can be thick or thin, dull or shiny, and the leaf posture could be upright, tilting, or drooping.

Like other monocotyledons, the upper and lower epidermis of *C. sinense* leaf is single layered and is covered with relatively thick waxy materials and it does not have a distinctive palisade and spongy mesophyll (Fig. 4.3). Stomata are found only on the lower epidermis at a comparative low density of 100–130 mm^{-2} and each stoma is covered with a cuticular ledge which is a common stomatal feature in leaves of epiphytic CAM orchids (Fig. 4.4). The presence of cuticular ledges reduces the diffusion of carbon dioxide and water molecules in and out of the leaves.[88]

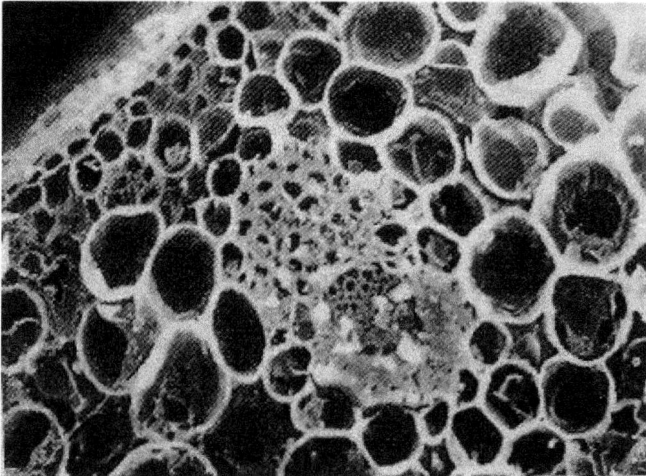

Fig. 4.3.   Scanning electron micrograph of a cross section of *C. sinense* leaf.

Fig. 4.4. Scanning electron micrographs of leaf stomata of *C. sinense*. (Left) Distribution of stomata; (Right) Stoma with a cuticular ledge, (SC).

Flowers

The flowers of the Chinese cymbidiums, like those of other orchids and monocotyledonous plants in general, have three sepals and three petals. The two lateral petals are usually equal in size and shape, whereas the bottom (middle) petal is different, resembles a lip, and is known as the labellum (Fig. 4.5). It functions as a landing platform for pollinators (e.g., bees, butterflies, moths). To a large extent, the beauty and diversity of orchid flowers are due to the different structures, colours, and forms of the labellum.

A unique feature of orchid flowers is the gynostemium (or column), which is a fusion of the male (stamen) and female (pistil) reproductive organs. The anther cap lies at the tip of the column, enclosing the pollinarium and the rostellum that lies beneath the pollinarium. Generally, the pollinarium consists of pollinia (mass of pollen), viscidium (a sticky disc), and stipe (a thin strip of tissue that connects the pollinia to the viscidium). The stigma, a shallow sticky depression in orchids, is located below the pollinia on the inner side of the column. It is connected to the ovary by the column and pollination is brought about by the deposition of the pollinarium into the stigma.[14]

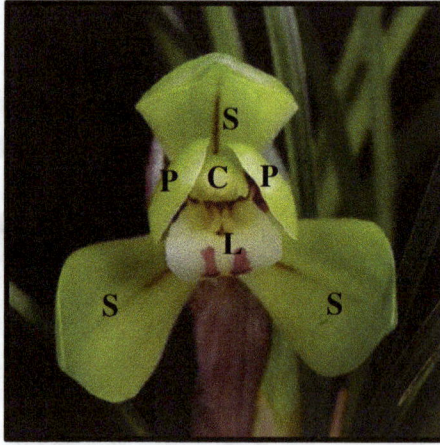

Fig. 4.5.   The floral parts of a Chinese *Cymbidium*. (**S**, sepal; **P**, petals; **L**, lip or labellum; **C**, column with anther cap).

As mentioned earlier, biologists have long been fascinated by the elaborate and intriguing orchid pollination mechanisms. The complex structure of orchid flower as an adaptive feature in ensuring an effective pollination is well documented.[2] A good example is the study of how the bee pollinates the orchid flower. After landing on the lip of the orchid flower, the bee will move slowly inwards in search of the nectary and its head will come in contact with the stigma. While the bee is moving out, its head will brush against the rostellum and the pollinia will stick onto its head. As this bee visits the next flower, it will deposit the pollinia onto the stigma of the second flower enabling fertilisation to take place.[2,7] Recent studies of *C. goeringii* and *C. kanran* showed the pollination of these two Chinese cymbidiums are odour guided and bee pollinated,[34,41] and the pollination mechanism is not different from that of other bee-pollinated orchids such as *C. insigne*.[7,41]

4.2. Flowers Art, Leaf Art, and Fragrance

Nowadays, when people appreciate the Chinese cymbidiums, they do not just pay attention to the colour and posture of the flowers and leaves but also to the connotations and artistic feel. Appreciating a Chinese *Cymbidium* is like appreciating a piece of artwork. This appreciation is often associated with Chinese art and culture, and it is known as "*Art of Lan*" (兰艺) which consists of "*Flower Art*" (花艺) and "*Leaf Art*" (叶艺).[65]

Flower art

The appreciation of the variations in flower shape, posture and colour of Chinese cymbidiums is commonly known as *Flower Art*. The flower shape generally refers to the posture of the sepals, petals, and labellum. According to the old Chinese floral terminology, petal and sepals in Chinese are both referred to as "ban" (瓣); the two lateral sepals are called "fu ban" (副瓣) and the middle sepal is called "zhu ban" (主瓣); while the petals are called "hua ban" (花瓣), which can be confusing. In compliance with the standard plant nomenclature, Wu[69] had suggested to reword sepal as "hua e" (花萼). Based on the different shapes of the sepals, Chinese cymbidiums can be divided into three main types: *Plum-shaped sepals, Lotus-shaped sepals,* and *Narcissus-shaped sepals*[65] (Fig. 4.6).

Plum-shaped sepals: The sepals are short, round, and curved slightly inwards, resembling that of a plum blossom. The *C. goeringii* 'Song Mei' (宋梅) and *C. faberi* 'Cheng Mei' (程梅) are widely regarded as typical representatives of the plum-shaped sepal flowers.

Lotus-shaped sepals: The sepals are broad and slightly elongated, with sharp or blunt tips, and narrowed at the base like the lotus

Fig. 4.6.   Variations in sepal types. (Left) Plum-shaped; (Middle) Lotus-shaped; (Right) Narcissus-shaped.

flower. Some examples are the *C. goeringii* 'Cui Ge He' (翠盖荷) and *C. goeringii* 'Lu Yun' (绿云).

Narcissus-shaped sepals: The sepals are relatively long and narrow, and the tip is slightly pointed or nearly sharp, with a narrow base. It looks very similar to the flower of a narcissus blossom. The *C. goeringii* 'Wang Zi' (汪字) and *C. faberi* 'Da Yi Pin' (大一品) have the typical Narcissus-shaped sepal flowers.

Butterfly-shaped type: This type of Chinese *Cymbidium* has undergone substantial changes in the floral parts with all of them becoming lip shaped. The lip is curled, with irregular cracks on the edges. In addition, there are patches or spots with different colours on the floral parts, resulting in gorgeous colouration. This floral form belongs to one of the oddly shaped types of Chinese cymbidiums. *C. ensifolium* 'Fu Shan Ji Die' (富山奇蝶) is one of the better-known varieties, with gorgeous flower colour, light floral fragrance and a long flowering period (Fig. 4.7).

The shape, colour, texture, venation and arrangement of sepals vary considerably in different varieties of Chinese cymbidiums. How the sepals are arranged is equally important to the growers and collectors because these features are also one of the major considerations for the appreciation and judging of Chinese cymbidiums.

C. goeringii 'Xiong Mao Rui Die' (熊猫蕊蝶)

C. goeringii 'Dong Fang Shi' (东方狮)

C. ensifolium 'Fu Shan Ji Die' (富山奇蝶)

C. Goeringii "Yu Hu Die" (余蝴蝶)

Fig. 4.7.   Exotic flower (variations in flower form).

The stretched posture of the lateral sepals can be divided into the following types[60]: (Fig. 4.8)

Flat sepals: The two lateral sepals are in a 180-degree straight line with each other, generally referred to as "*flat sepals*" or "*one-line sepals*" (平肩). This is regarded as a highly valued feature of Chinese cymbidiums.

Flying sepals: The two lateral sepals face upwards and resemble a bird spreading its wings upwards and taking flight (飞肩). This posture is regarded as a graceful feature of the Chinese cymbidiums.

Fig. 4.8. Types of sepal arrangement in Chinese *Cymbidium* flowers. (Left) Flat-sepals; (Middle) Flying-sepals; (Right) Drooping-sepals.

Drooping sepals: This less desirable feature is when the sepals on both sides of the flower are slightly curved downwards (落肩), resembling the posture of a bird gathering its wings before landing.

The first two types of sepal arrangement are more popular, because they represent a kind of positive, motivating and lively feeling, whereas the posture of the drooping sepals depicts despair and sadness.

The colour of Chinese *Cymbidium* flowers is of great significance too. For example, "*Su xin lan*" (素心兰) is unique (Fig. 4.9). "*Su xin*" refers to a flower with its sepals, petals, and pedicel being one pure colour. The literal translation of the Chinese words is "*Pure of heart*", thus meaning purity, simplicity, and modesty. It could be white, greenish-white, yellow, pink, orange, or without variegation, stripes, or spots. This unique flower colouration is a result of genetic variation. Due to its purity in colouration, rarity and pleasant fragrance, *Su xin lan* is usually expensive and has been well sought after by orchid lovers in China since ancient times.

Su xin lan is also popularly grown by ethnic minority groups in China. For instance, *C. ensifolium* 'Bian Cheng Gong Su' (边城贡素), with its dignified, greenish-white flowers and sweet fragrance, was once given as a tribute to the emperor of the Late Qing Dynasty by the Miao King of Xiangxi (湘西苗王).[61] Some other common representatives of *Su xin lan* varieties are *C. ensifolium*

Su Xin Lan, *C. ensifolium* 'Tie Gu Su'
(铁骨素)

C. ensifolium 'Da Tang Gong Fen'
(大唐宫粉)

Fig. 4.9.   Su Xin Lan.

'Tie Gu Su' (铁骨素), *C. ensifolium* 'Long Yan Su' (龙岩素) and
C. kanran 'Ju Hong Su' (橘红素).

'Tie Gu Su' (Fig. 4.9) is a famous *Su xin* variety of *C. ensifolium*
which is highly treasured by many orchid lovers. The dark green and
strong upright leaves are tough in texture like iron wire, and the
pedicel is thin, erect and has strong supporting force like iron tendons.
Its fragrance is mellow; it is fast growing and easy to maintain. With
its reasonable price, it is a very suitable plant for home cultivation.

Leaf art

Leaf Art refers to the artistic appreciation of the variation in the
structure, shape, colour, posture, texture and variegation of Chinese
Cymbidium leaves.[58,65] The design and pattern of the artwork in
Chinese *Cymbidium* leaves are diverse and complex. These patterns are
a result of genetic mutation hence are unpredictable and cannot be
reproduced, making each leaf unique and rare. Colour spots or strips
often appear on the leaf surface. Spots can be scattered or in patches

and white or yellow patches can appear at the tip, along the edges or in the middle of a leaf. Common leaf variegations include spotty stripes, saturated mid-leaf spots, snake-skin spotting, cloudy fountain marks, transparent middle veins and crystalline marks.[58,65] The shape of Chinese *Cymbidium* leaves can also be remarkably diverse, with some resembling that of a goose head, crouching dragon, and dancing phoenix, while others are twisted and have wrinkled surfaces (Fig. 4.10). The graceful and diverse types of leaf variations undoubtedly add colourful diversity to the Chinese cymbidiums.

Fig. 4.10.   Leaf variegation in Chinese cymbidiums.

Fig. 4.10. (*Continued*)

The relative amount of different plant pigments present in the leaf determines the intensity and pattern of colouration of leaves. Chlorophylls and carotenoids are pigments present in the leaf all year long and they determine the colour of a leaf. The variegated leaf, for example, is a result of an uneven distribution of these pigments.

In recent years, leaf art has attracted great attention amongst Chinese *Cymbidium* lovers. There is a noticeable change in the perception of beauty and the overall appreciation of the Chinese cymbidiums. Leaf art has now become a more fascinating, mysterious, and colourful artwork than floral art.

In China, large commercial nurseries devoted to growing exclusively the Chinese cymbidiums with diverse leaf forms have emerged in recent years. *C. sinense* having long, broad and dark shiny leaves is the preferred choice. Orchid growers will separate a pot of 3- to 4-year-old *Cymbidium* plants into three age groups: young, middle and old. The middle and young ones are placed together in one new pot and are ready for sale. The old one with healthy pseudobulbs and roots is potted separately for growing new shoots. During repotting, part of the leaves are removed to allow more nutrients for the young growing leaves. It is also a common practice to remove the young flower reaching 5–8 cm in length to encourage more rigorous leaf growth. The right proportion of nitrogen and magnesium in fertilizer is crucial for healthy leaf growth.

The art of appreciation of the Chinese cymbidiums also applies to the dwarf varieties. In fact, all the five Chinese *Cymbidium* species mentioned in Chapter 3 have dwarf varieties.[71] These dwarf varieties have an appealing plant form and remarkably diverse leaf variations, making them very attractive and well liked by Chinese orchid lovers. Dwarf Chinese cymbidiums have long existed in nature. However, in the past, some people thought that the dwarf varieties were too short and insignificant, and were unwilling to grow them. However, they soon realised the ornamental value of these plants. There are three main factors that contributed to the change: (1). The plants are noticeably short but well balanced, presenting a unique aesthetic three-dimensional beauty. (2). The leaves are short, thick, dark green, shiny, and have attractive leaf forms. (3). The roots are relatively large and sturdy, giving it a strong and fresh appearance. As such, it is now a popular trend to place the dwarf orchid variety in the living room.

C. sinense has the greatest number of dwarf varieties followed by that of *C. ensifolium, C. faberi,* and *C. kanran.*[71] Two of the highly treasured dwarf varieties are *C. goeringii* 'Lu Yun' (绿云) and *C. sinense* 'Da Mo' (达摩). *C. goeringii* 'Lu Yun' is a rare and famous dwarf variety that has the reputation of being the "Queen of *C. goeringii*". It was listed as a "gem" in the "*Lan Hui Tong Xin Lu*" (兰蕙同心录) of the Qing Dynasty. Its leaves are dark green, short, broad and thick, about 15–23 cm long and 1–1.2 cm wide. The lotus-shaped sepal-type flowers are beautiful, with a strong fragrance and are well treasured by orchid collectors (Fig. 4.11).

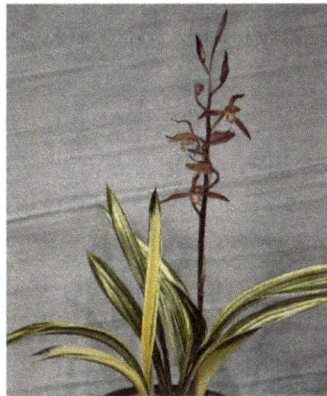

C. goeringii 'Lu Yun' (绿云)	*C. sinense* 'Da Mo' (达摩)

Fig. 4.11.   Dwarf varieties of Chinese cymbidiums.

C. sinense 'Da Mo' is called the "*King of Leaf Art*" as it is rich in variants and its leaf art continues to evolve. It was first found growing wild in Hualien, Taiwan. The leaves are short, thick, silky and wide, and the middle part of the leaf is often slightly twisted. The flower stalk is tall and much above the leaves, and the overall plant shape is graceful and conspicuous. As the leaf art of this dwarf species varies greatly and is unpredictable, it has a very high ornamental and commercial value. Thus, it is a precious orchid variety well sought after by the *Cymbidium* orchid community (Fig. 4.11).

Flower fragrance

Another unique characteristic of Chinese cymbidiums is its flower fragrance. It is an equally important aspect in the judging and appreciation of Chinese cymbidiums. It is regarded as the "*soul*" of Chinese cymbidiums and has earned the reputation of being referred to as "*King of Fragrance*" by Confucius. The intensity of fragrance of different species/varieties of Chinese cymbidiums growing in different environments and regions varies; its fragrance can be broadly classified into 4 groups: mellow, delicate, sweet and strong.[57]

C. goeringii: The "delicate fragrance" (清香) of this Chinese *Cymbidium* is widely accepted as the purest and the best of all Chinese cymbidiums. Its scent intensity depends on the location where it grows. For example, the *C. goeringii* growing in southeast China has a strong scent, while those grown in the west and southwest China have a weaker smell. As for *C. goeringii* growing in the northern region of China, most of the flowers are scentless or only have a very faint fragrance.[61] *C. goeringii* 'Song Mei' (宋梅), which appeared as early as the Qianlong reign in the Qing Dynasty, is revered as one of the best amongst the *C. goeringii* varieties. Its flower shape is magnificent and its fragrance is unique and delicate. It is loved and pursued by many orchid lovers since ancient times.

C. faberi: Most *C. faberi* have a strong fragrance (浓香). This is attributed to the relatively large numbers of the flowers (per stalk) blooming at the same time. Moreover, the flowering season of *C. faberi* is usually in late March and April when the air temperature is high, and the production of fragrance is enhanced. However, there are some who feel that the scent can be a bit too overwhelming, and it smells best from a distance away.

C. ensifolium: There are some varieties of *C. ensifolium* which have single-colored flowers with sweet and pure fragrance. For example, the *C. ensifolium* 'Yong Fu Su' (永福素), a *Su xin lan*, the flowers are large with a long flowering period, and has unique and refreshing floral fragrance. *C. ensifolium* 'Yin Bian Da Gong' (银边大贡) also a *Su xin lan*, is a popular household flower in Fujian and Guangdong. The stems are tall and straight; the silver-edged leaves are arched and semicircular; the light color flowers are numerous with a sweet and mellow fragrance.

C. sinense: Most *C. sinense* have a faint fragrance (淡香) compared with other Chinese cymbidiums. *C. sinense* 'Jin Bian Mo Lan' (金边墨兰) is one of the more fragrant varieties.

C. kanran: *C. kanran* blooms in winter, the flowering period is from October to January of the following year, and most varieties of *C. kanran* produce a light and delicate fragrance. Based on the leaf form, it is divided into "*Broad-leaved Han Lan*" and "*Thin-leaved Han Lan*". The former does not have strict environmental requirements such as sunlight or temperature difference for fragrance production. However, the *thin-leaved Han Lan* is sensitive to the influence of the cultivation environment and nutrient requirements. It requires stricter conditions to produce its fragrance.[57]

Flowers of *C. kanran* generally emit fragrance at least three days after blooming. The floral scent is more intense at a low temperature of 5–10°C and with a large temperature difference. For instance, the emittance of scent from *Thin-leaved Han Lan* requires a low temperature of 8°C or below. Therefore, they are more suitable for growing in Zhejiang, Guangxi, Hunan, Hubei, northern Fujian and Jiangxi where the temperature is relatively low in winter. The scent of the "*Thin-leaved Han Lan*" is generally more delicate than that of the "*Broad-leaved Han Lan*".[48]

When the flowers of *C. kanran* are in full bloom during winter, the cool breeze will bring along a delicate fragrance, giving one a peaceful and refreshing feeling. This unique type of fragrance is known as "*Cold Fragrance*" (冷香). There is a report of another interesting type of fragrance produced by some varieties of *C. kanran* which is known as "*Remnant Fragrance*" (遗香). The flower will continue to emit a faint fragrance even when it has already started to wither, provided that the column and lip are still intact.[57]

In short, different Chinese cymbidiums have different floral scents. Therefore, the "pleasant and delicate fragrance" of *C. goeringii*, "strong fragrance" of *C. faberi*, "light fragrance" of *C. sinense*, "pure and fresh fragrance" of *Su xin lan*, and the "cold and remnant fragrance" of the *C. kanran* are all unique and distinctive. These interesting aspects of floral scent have greatly enriched the unique characteristics of Chinese cymbidiums.

Research has shown that the fragrance of the orchid flower comes from the column.[88] The intensity increases with flower development and decreases following the senescence of the flower. The senescence in orchid flowers can be induced either by pollination or removal of the pollinium. The composition of the fragrance is complex. Xu (2019)[37] has reported that methyl jasmonate is one of the main volatile compounds produced by Chinese *Cymbidium* flowers. In fact, there are more than 56 chemical compounds identified in the fragrance produced by *C. sinense* flower.[88] The interaction of these volatile compounds makes it difficult for the detection and analysis by the receptors of the human olfactory. The composition of the floral fragrance is species specific, and it is important for the classification of orchid species as well as for pollinators.[7] Difficulties in the differentiation of scent profiles have been a major challenge for Chinese *Cymbidium* breeders.

However, in recent years, considerable advances were made in the analysis of orchid fragrances including the use of an "electronic nose". It consists of an array of sensors with each having a different sensitivity to a wide range of chemical compounds.[20]

4.3. Physiology and Biochemistry

The horticultural practices of growing Chinese cymbidiums in the past were primarily based on experience. Early publications focussed mainly on distribution, taxonomy, and cultural practices. Optimisation of the growth and development of the Chinese cymbidiums requires a good knowledge of their physiology and biochemistry. A mini review on the physiology of Chinese cymbidiums was published in 1997 and this was followed by a more comprehensive review nine years later.[27,88]

Over the years, there have been considerable advances made in the understanding of the physiology and biochemistry of Chinese cymbidiums. These include carbon metabolism, mass clonal propagation, orchid mycorrhiza, molecular biology, flowering, biodiversity and conservation.

Seed germination and mass clonal propagation

After pollination and fertilisation, the ovary will develop into a fruit capsule (seed pod) (Fig. 4.12). The days to maturity of the capsule depends on each orchid species. For *C. sinense* and *C. ensifolium*, the days to maturity would be between 150 and 190 days, respectively. Germination of the seeds is best when the capsule is at 80% maturity and has turned brown but not yet split open.[88] Like other orchids, Chinese *Cymbidium* seeds are very minute, dust like, and spindle shaped with an average length of about 1 mm, a width of 0.1–0.2 mm and a weight of 0.3–1.4 μg. Each capsule may have as many as

Fig. 4.12.   Young seed pod of a Chinese *Cymbidium.*

millions of seeds and is enclosed by a single layer of the seed coat (Figs. 4.13 and 4.14) with the absence of the endosperm. Apart from the small starch grain within the protoplasts, there are no other carbohydrate reserves in the orchid seeds.

In nature, orchid seed germination is possible only following colonisation of the seed by specific fungi and the establishment of a symbiotic relationship between the two. This is known as "*symbiotic germination*". This process starts when the fungus hyphae enter the host (orchid) through either the roots or the seed through an opening in the seed coat. However, there is no evidence of a specific mechanism that attracts the fungus to the seed, but it is known that orchid seed germination would only take place if the right fungi were present. The fungi are known to provide food and other compounds like vitamins to the orchid host. In return, the host provides amino acids and other compounds necessary for fungal growth.

Fig. 4.13. Mature seed pod of Chinese *Cymbidium*. (Left) Cut-open seed pod; (Right) Minute seeds.

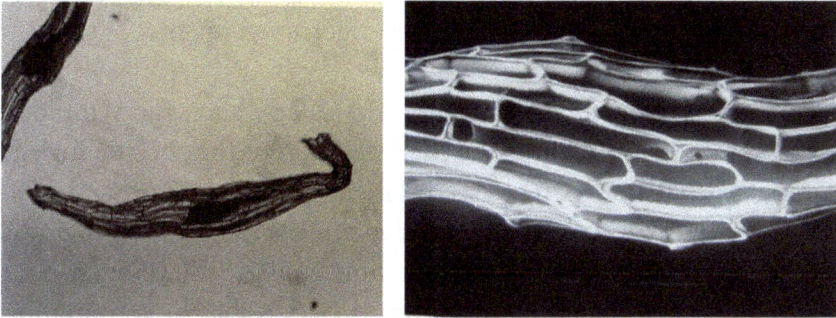

Fig. 4.14. Seeds of *C. sinense*. (Left) Light microscope photograph of a minute seed; (Right) Scanning electron micrograph of the surface of a mature seed.

Fungi in the roots of orchids were apparently seen, but their beneficial effect was not recognised until 1824.[2] Orchid seeds were first germinated horticulturally in England in the mid-1800s and the role of fungi in orchid seed germination was first discovered in 1899 by Noel Bernard in France. A medium for symbiotic seed germination was formulated only as late as the 1920s in the USA.[2] It would be of interest to know when symbiotic seed germination of orchids began in China. It is believed that the use of mycorrhizal fungi to stimulate the germination of the seeds of valuable medicinal orchids was carried out back in the 1980s in China.[88] Active research on the complexity of the association between Chinese cymbidiums and endophytic fungi began in the 1990s.

Fungi are also found mainly in the cortex of roots of Chinese cymbidiums. Endophytic fungi (fungi inside the plant) including *Rhizoctonia* and *Mycena* from roots of the wild adult plant of *C. ensifolium* and *C. sinense* have been successfully isolated and studied. Endophytic fungi in the root of the *C. goeringii* were found to alter nitrogen acquisition by the roots and to enhance its growth.[36] In another study, the endophytic fungi in the roots of mature *C. goeringii* were isolated. They were homogenised into a liquid fungal preparation, which was used as an elicitor for seed germination. Such an approach increased the seed germination of *C. goeringii* from 15% to 30%.[9]

One difficulty in the development of methods for symbiotic orchid seed germination is the identification of the specific fungus compatible with each orchid seed. The American plant physiologist, Lewis Knudson, discovered that the main function of the fungi was to provide orchid embryos with soluble sugars, which served as a source of energy. Knudson's discovery made possible the germination of orchid seeds in the absence of fungi (i.e., *asymbiotic germination*) through the inclusion of sucrose in the culture medium to give the orchid seeds the necessary sugar required. Subsequently, several culture media were formulated for the asymbiotic seed germination of Chinese cymbidiums and other orchids. These included the Knudson, Murashige and Skoog, and Vacin and Went media.[1,65]

The successful formulation of orchid seed germination and tissue culture media greatly enhanced studies of the germination of both mature and immature seed and explant culture of Chinese cymbidiums.[5,38] The immature seeds of *C. faberi*, for example, were reported to form rhizomes four months after they were placed on a standard culture medium. When rhizomes were placed on a medium containing plant hormone, shoot growth was induced and plantlets were produced.

Vegetative multiplication of Chinese cymbidiums is made possible by simply dividing large pseudobulb clusters into smaller ones and planting them separately. It is a simple but slow process for the propagation of orchids. Rapid mass clonal micropropagation of plants including Chinese cymbidiums has since been developed to meet the fast-growing consumers' demand for orchids.[31,39]

To start mass clonal micropropagation of Chinese orchids, a shoot apex or young bud is excised and placed in a liquid or on a solid culture medium containing appropriate components. If a liquid medium is used, the flask should be placed under illumination on a shaker which will ensure aeration of the medium. The explant (shoot tip) of *C. goeringii*, for example, will swell after two weeks and will further develop into masses of cells, known as protocorms, in a month. When placed in suitable culture medium, (e.g., Murashige and Skoog medium with coconut water), the protocorms will develop into numerous tiny plantlets, each of which will give rise to a plant identical to the mother plant[88] (Figs. 4.15–4.18).

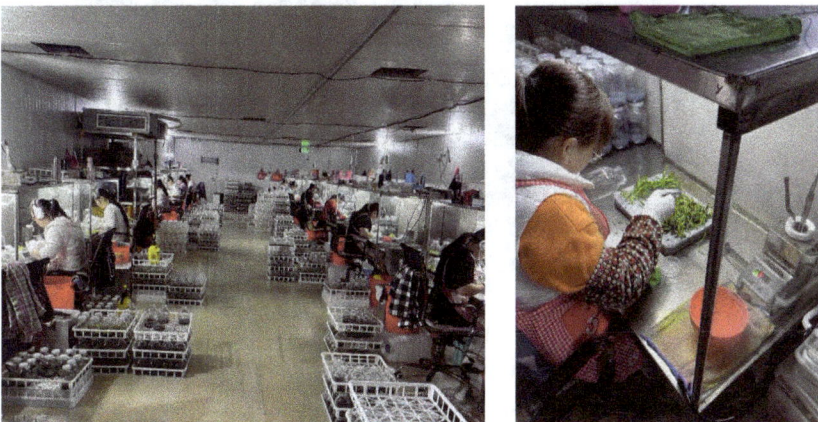

Fig. 4.15.   Commercial Chinese *Cymbidium* tissue culture laboratory.

Fig. 4.16.   Stages in tissue culture of Chinese *Cymbidium*. (a) Shaker for agitating orchid tissue in liquid medium; (b) Initial stage showing protocorms; (c) Placing protocorms into solid medium; (d) Development of plantlets from protocorms in solid medium; (e) Developing plantlets; (f) Plantlets ready for transfer to nursery.

Fig. 4.17.   Hardening of plantlets in a commercial greenhouse.

Fig. 4.18.   Commercial cultivation of Chinese cymbidiums.

Mass clonal propagation of Chinese cymbidiums has been actively pursued in China. Improved tissue culture methods using orchid roots, leaves, flowers buds, and inflorescences have been adopted. Different media have been formulated to obtain optimal growth and differentiation of protocorms of the commercially important Chinese *Cymbidium* species. However, there are still many problems associated with the commercial production of *Cymbidium* plantlets — low multiplication rate, poor rooting and high mortality rate during acclimatisation. Recently, there has been a renewed interest in improving *in-vitro* culture conditions by the optimisation of environmental factors that include light, gaseous environment, temperature and humidity.[32,88]

Photosynthesis, respiration and translocation

Extensive research has been carried out to understand how carbon dioxide is biochemically fixed by green leaves. There are at least three different pathways for carbon fixation — C_3, C_4, and CAM pathways. Plants can be divided into three major groups with respect to the biochemistry of the carbon dioxide fixation pathway[14]:

(1) The first group is generally referred to as the C_3 plant. The fixation of CO_2 is carried out via the Calvin cycle (or C_3 cycle) and the first stable product of photosynthesis is a three-carbon compound. For example, sunflowers and spinach are C_3 plants.

(2) The second group is the C_4 plants. They fix CO_2 through the C_4 pathway, and the first stable product of photosynthesis is a four-carbon compound. Corn and sugarcane are examples of C_4 plants.

(3) The third group is known as *Crassulacean Acid Metabolism* (CAM) plants. Fixation of CO_2 is through the CAM Pathway. The first stable product of CAM photosynthesis is also a four-carbon compound. Fixation of CO_2 is carried out mainly in the dark. Cactus and pineapple are common CAM plants.

Photosynthetic rates and light saturation are different for these three plant groups. C_4 plants have the highest rate of photosynthesis and prefer full sunlight. Photosynthesis of C_3

plants saturates at about 1/4 to 1/3 of full sunlight. C_3 plants have a rate of photosynthesis about 1/4 to 1/2 that of C_4 plants. CAM plants have the lowest rate of photosynthesis and saturate at a much lower light intensity than the C_3 plants. Hence, C_4 plants have the highest rate of photosynthesis followed by C_3 plants; the CAM plants have the lowest rate of photosynthesis. The rate of growth is, therefore, fastest in C_4 plants. There are C_3 and CAM orchids but no C_4 orchids.[14]

The leaves of the orchids can be divided into two groups: thin-leaved and thick-leaved. The thickness of thin-leaved orchids ranges from 0.2 mm to 0.65 mm, whereas that of thick-leaved orchids is 1.5–2.5 mm. Thin-leaved terrestrial orchids like *Arundina* are C_3 plants; thick-leaved epiphytic orchids like *Phalaenopsis* are CAM plants. Terrestrial Chinese cymbidiums are thin-leaved C_3 plants, and the rates of photosynthesis and light saturation point are comparable to those reported for other C_3 thin-leaved shade-loving orchids such as *Oncidium* Golden Shower. The photosynthetic rate of Chinese cymbidiums is generally low and is about 1/5 that of soybeans.

Photosynthesis and respiration are two very important metabolic processes in plants. Respiration, in breaking down the carbohydrates formed during photosynthesis, produces the necessary food and energy required for plant growth and maintenance of living tissues. Oxygen is consumed and carbon dioxide is released during respiration. Starch, sugar, fats, organic acid, and under some conditions even proteins can serve as respiratory substrates.

The respiration of tropical orchids has been studied and reviewed.[11,14] By comparison, the respiration of seeds, roots, and flowers has received much attention. There has been considerably more work done in the respiration of germinating orchid seeds. The rate of flower respiration varies with orchid species/hybrids, age, and environmental factors such as temperature. An inverse relationship between respiration rates and flower longevity has been reported. The short-lived orchid flower has a higher respiration rate as compared to a long-lived flower. There is an increase in respiration rate following pollination. A change in respiratory rates (respiratory drift) has been observed during orchid flower development. Pan and Ye[88] have studied the respiration of *C. sinense* and other Chinese cymbidiums. The optimum temperature for leaf respiration of *C. sinense* is 30°C. When the root is kept at a temperature higher than 30°C for one hour, the respiratory rate decreases sharply.

Extensive research has also been carried out on the translocation and distribution of photosynthetic products in tropical orchids.[14] It is an important aspect of how plants regulate growth and development. Essentially, the transport and distribution of photosynthetic products follow a supply and demand (source–sink) relationship. The source (leaf) and sink (roots or other actively growing organs) relationship changes with plant development. During vegetative growth, actively growing root tips or pseudobulbs are the active sinks in attracting most of the photosynthetic products. Following flower induction, the developing inflorescence becomes the major sink. The same has also been reported in *C. sinense*.[88]

Orchid pseudobulbs are known to function as water and food storage organs. The importance of pseudobulbs in the remobilisation of food for the growth of Chinese cymbidiums is well illustrated by a fascinating observation made by an orchid owner of a large orchid nursery in Liancheng, Fujian, China. He noticed that one of his Chinese *Cymbidium* that was left at a corner in his nursery unattended for months was still growing strong and healthily. Upon careful observation, he observed that the pseudobulbs were connected in a row, by what he called the "*umbilical cord*" (Fig. 4.19). He reasoned that the pseudobulbs were interdependent and supplying food and water to many parts of the plant through this umbilical cord.[84]

The remarkable ability for remobilisation of photosynthetically fixed products or storage food from old and current pseudobulbs to

Fig. 4.19. Showing the "umbilical cord" between pseudobulbs.[84]

active growth centres has been well studied in *Oncidium* Golden Shower, a sympodial orchid.[14] Direct evidence to demonstrate that the shoots (pseudobulbs) of *Oncidium* Golden Shower are physiologically connected and interdependent for photosynthetic products is obtained from experiments using radioactive carbon ^{14}C as a tracer. Experimental evidence indicates there is a movement of the ^{14}C labelled photosynthetic products in the leaf of the back shoots to the active sink (such as developing flower bud) on the current shoot. A similar pattern of the distribution and remobilisation of photosynthetic products of *C. sinense*, a sympodial orchid, has also been reported.[88]

Water relation and mineral nutrition

Most orchids are either epiphytic or terrestrial. Epiphytic orchids such as *Cattleya* and *Bulbophyllum* are found normally on treetops and face a dry environment. They can survive under water-stressed conditions, by having highly evolved structures that enable them to conserve water. These include prominent pseudobulbs, thick leaves, xerophytic stomatal structures, and fleshy roots with velamen. Chinese cymbidiums are terrestrials, but they also possess some of these special adaptive features (See Chapter 4.1. Structure and Characteristics of the Plant Parts). A study carried out by Pan and Ye[88] has demonstrated that Chinese *Cymbidium* can withstand severe drought. Even when the soil water content is reduced to 12% of the maximum holding capacity after 42 days of drought, the water content of roots, leaves, and pseudobulbs of *C. sinense* remains at 63–70%. Leaves of different ages respond differently to drought.

Like other plants, Chinese cymbidiums require essential mineral elements for normal growth. Some essential elements are required in larger quantities (macro-elements), whereas others (micro-elements) are needed in trace amounts. Deficiencies of an element may develop when its concentration drops below a level necessary for optimal growth. The concentrations of macro- and micro-elements in tissues of most plants (including orchids) have been studied, and the levels for various elements considered adequate for orchid growth are well documented.[28] The composition of mineral content in plant tissues, typically leaves, is determined by tissues analysis. There exists a relationship between plant growth and the mineral content of plant tissues. When the nutrient content in tissues is low, growth is retarded.

The three macro-elements essential for normal plant growth are nitrogen, phosphorus, and potassium. These three macro-elements, each have their physiological functions and together they are responsible for various metabolic processes that ensure the normal growth of plants.

Nitrogen (N): Nitrogen is associated with many plant cell components such as amino acids and nucleic acids. It is a component of chlorophyll that is necessary for photosynthesis. The colour of leaves will turn yellowish (chlorosis) when nitrogen is deficient. In severe cases, the leaves will become completely yellow and fall off; younger leaves will remain green for a longer period because they receive soluble nitrogen transported from the older leaves. However, when nitrogen is in excess, the plant has abundant dark green foliage but will not flower.

Phosphorous (P): Phosphorous is an essential component of many compounds present in the cell. This includes the sugar phosphates involved in photosynthesis, respiration, and phospholipids in cell membranes. Phosphorous is also part of the nucleotides used in the energy metabolism of plants. It stimulates root growth, promotes flowering, and increases both cold and drought resistance in plants. Phosphorous deficiency symptoms occur first in mature leaves because phosphate is easily redistributed from older leaves to the young expanding leaves.

Potassium (K): Potassium and phosphorous have some overlapping roles. They both stimulate root growth and stem development and improve the ability of plant resistance to diseases, cold and drought. Potassium plays an important role as an activator of many enzymes that are essential for photosynthesis and respiration, and of those involved in starch and protein synthesis. When potassium is present in insufficient amounts, the plants are weak and are unable to stand upright, and the flower colour is not bright. The first symptom of potassium deficiency is chlorosis (yellowing) at the leaf margin which subsequently develops into necrosis (death) of the leaf. Similar to phosphorous, potassium deficiency symptoms also first appear in the older, mature leaves.

Orchids in general may take longer to show mineral deficiency. The slow development of mineral deficiency in epiphytic orchids is related to their remarkable ability to remobilise minerals from other leaves and storage organs such as pseudobulbs.[13,14] The uptake rate of minerals by most orchid roots is relatively slow in comparison to other plants such as corn. The slow rate of mineral uptake is attributed either to the barrier encountered at the interface between exodermis and cortex or the limited supply of respiratory substrates to the roots. A slow rate of uptake of minerals by roots of *C. sinense* has also been reported. Detailed studies of mineral nutrition of *C. sinense* have been carried out by Pan and Ye (2006).[88] Generally, nitrate nitrogen is a better source for growth. However, excellent leaf and root growth were observed when ammonium nitrogen and nitrate nitrogen were applied in an appropriate concentration as the nitrogenous sources.

Growth and development

Flowering is an important part of plant growth. Through flowering, sexual reproduction can be brought about and new hybrids can be produced. The process of flowering in orchids as in all other angiosperms consists of two parts: flower induction (flower initiation) and floral development. Induction of flower initiation is affected by genetic, environmental, and physiological factors. Following induction, floral buds and their subsequent growth depend on the supply of foods from various sources.[2]

The three main factors affecting the induction of flowering are juvenility, vernalisation (response to low temperature), and photoperiodism (response to day length). Juvenility refers to the early phase of plant growth during which flowering cannot be induced by any treatment. The juvenility phase of orchids varies; the average time is between 2 and 3 years, and for *C. sinense* it is about 4 years.[2,88] Under favourable environmental conditions, the flower bud of *C. sinense* emerges at the base of the pseudobulb. The flowers usually bloom either in January or February. Differentiation of the flower bud starts between August and November the year before, and it takes two to two and a half months for the flower to be fully developed. The flower bud grows rapidly when it is placed under a 20/15°C regime.

Hybridisation is a common practice used to produce new hybrids. Keeping pollinia (masses of pollen) fresh and healthy is especially important for the success of a hybridisation programme. This is particularly so when the pollinia are collected from plants growing at a different location. It has been reported that pollinia of *C. sinense* remain fresh and fertile when they are kept in a closed vial placed in a desiccator and stored at 4–6°C for as long as 280 days. Exceptionally good fertilisation rates (almost 100%) have been obtained with these pollinia following artificial pollination.[88]

Chapter 5

Cultivation of Chinese *Cymbidium*

The cultivation of Chinese cymbidiums in China can be traced back to more than a thousand years ago. The practices used in the past would certainly be of interest to present orchid growers and researchers. We will first summarise the horticultural practices of Chinese cymbidiums in the past, followed by a description and discussion of the current cultivation practices.

5.1. Past Horticultural Practices

A general account of the literature on the early cultivation of Chinese cymbidiums has been discussed in Chapter One. We will now examine the past horticultural practices of Chinese cymbidiums as described in four early pieces of literatures published during the Song period (960–1279). These are *Jing Zhang Lan Pu* (1233), *Wang Shi Lan Pu* (1247), *Lan I*, and *Lan Pu Ao Fa* (year of publication unknown) and are often regarded as the most important documents in providing insights into the past horticultural practice of Chinese cymbidiums.

Planting and repotting

Propagation is an important part in cultivation of Chinese cymbidiums. People in the past had already noticed that Chinese cymbidiums liked to grow in groups and the plants did not grow well when they were separated from their clumps or cultivated as a single plant.[56,78,85] However, the division of clumps is necessary when they outgrow the container in which they are planted. To repot the plants, it is recommended to break the pot, carefully remove the damp medium, and cut away the rotten roots. Alternatively, one can place the pot in water, loosen the potting mix, and get the plant out slowly to trim the roots. Three to four shoots can be put together in one pot, but the oldest shoot should be placed in the middle.[59,78,85]

Plants must initially be planted slightly deeper and then slowly raised higher to prevent water being trapped amongst the roots. The bottom of the pot needs to be filled with potting material that allows water to flow easily. It is also recommended to half-fill the pot with broken bricks and cover the top surface with fine sand. The plant can be watered with pond water or rainwater, and it is best to refrain from watering and applying fertilizers for two weeks after transplanting.[56,85]

Division of the plant is best carried out in late autumn or early winter when growth is slow, or when the buds are dormant. To conserve the "*qi*" (气; energy, vital force forming part of any living entity) of the newly planted Chinese *Cymbidium*, it is necessary to remove the first flower produced in the second year after planting. In addition, it is important to ensure that leaves and roots are growing well as this will ensure good flowering.[56,85]

When collecting Chinese cymbidiums from the wild, one should collect some of the soil or debris along with the plant as this is important for the initial establishment of the wild orchids.[85] It is also the preferred potting medium and can be mixed with dry animal manure, preferably that of geese or goats.[56,85] The recommendation of using only dried animal manure is worth noting. The drying of animal manure before use is commonly practised today to remove ammonia which is toxic to the plant.

Burnt soil, or as some call it, "golden soil" (金泥), is commonly used as a potting medium and is prepared by burning good-quality mountain soil under a small fire. Pig or cow bone ash can also be used as fertilizer and the preferred ratio of burnt soil to animal bone ash is 4:3. Fresh soybeans or cooked beans can also be used, but they must be soaked in boiling water before storage and usage. The organic fertilizer should be dried, and the frequency of fertilizer application would depend on the growth of leaves and roots, preferably not more than 2 to 3 times a month. It is recommended to apply fertilizer once a month in October and in November when newly emerging buds are 12–13 cm in length right after flowering. However, one should not apply fertilizer when the bud has just emerged.[85]

Watering

Watering regimes recommended for Chinese cymbidium cultivation in olden days depended on the prevailing environmental conditions, plant type, and how well the plants were growing. The best water source is the early morning mist, particularly when the plants are kept outdoors overnight. Rainwater can be used if the mist is not available. Cold well water is not recommended, particularly in the early spring.[59,85] During the rainy season from April to May, watering should be reduced or stopped. In July, the amount of water used should depend on the potting mix and the condition of the plants. In addition, when the weather is still relatively warm in early autumn and the potting medium is damp, watering should be carefully controlled. In September, care should be taken to avoid overnight freezing injuries.[85] Watering the plants with water used for cleaning fish can protect the plants from being injured by ice formation. When the roots are young, one needs to refrain from heavy fertilisation and watering. Early Chinese *Cymbidium* growers were advised not to overwater their plants and to treat each species/variety differently with regard to watering, fertilizer application and selection of potting mixture.[78,85]

Pest control

The major pests of Chinese cymbidiums in early China were ants, rats, earthworms, spiders, lice, and snails. Tall flower stands with wire meshes were set up to prevent rat eating roots of the plants. Ants do not cause great harm to the orchids, but they do take away the precious nectar from the flowers. To control earthworms, one can soak the pot in water, as this would drown the worms. Human urine can also be used to drive away worms. Spiders form webs on leaf surfaces and cannot be seen by the naked eye, but they suck away the *qi* (energy) of leaves, causing them to turn yellow and wilt. Garlic can be mixed with water and brushed onto the leaves to remove "ji shi" (蟣蝨; lice).[56,78,85] Other measures suggested for pest control include using a painting brush to apply sesame oil, "yu xing shui" (鱼腥水; fishy water), or "zhu bang tang" (煮蚌汤; boiled mussel soup) onto the leaves surface. The use of ashes from burnt charcoal has also been suggested to control pests. It is also important that infested plants be isolated to prevent spreading. Poor ventilation invites pests. After being exposed to rain, plants should be covered when the sun is out. White spots will develop when wet leaves are exposed to strong sun.[59,85]

One can observe the beginning of the use of natural pesticides and repellents in ancient Chinese horticultural practices. The proposed method for controlling minute lice (蚤) by using oil or an oil-based mixture is also interesting. Oil controls scale insects and other minute pests mainly by suffocation. A good example is the current use of white summer oil to control scale insects. The minute lice mentioned in the early publications could likely have included mealybugs and other pests such as scale insects.[64] The presence of scale insects is consistent with the ancient practice of scraping scale insects off by using thin bamboo pieces.[56,85]

The utilisation of plant extracts such as garlic and Chinese honey locust (*Gleditsia sinensis*) to control orchid pests is also interesting.[85] *Gleditsia sinensis*, also known as "*Zao jiao*" (皂角), is a herb that has been in use for traditional medical treatment in China for a long time. It is listed in *Ben Cao Gang Mu* and is used as a medicinal herb for the treatment of inflammatory and skin diseases as well as swelling. It is slightly toxic and has been shown to inhibit bacteria growth. It has also been suggested for use in the control of maggots and flies.

It is both fascinating and refreshing to read about Chinese *Cymbidium* cultivation in ancient China. Not only do we derive good insights into its cultivation in those days in China but we also learn about the passion that the ancient Chinese orchid growers had for Chinese cymbidiums. This passion resulted in the development of useful cultivation practices. When reviewing the old literature on Chinese cymbidium cultivation in ancient China, one is impressed by the sound knowledge growers had regarding the growth and physiology of orchids. It is therefore not surprising that many of the age-old practices are still in use today.

5.2. Present Horticultural Practices

Chinese cymbidiums originally grew wild in mountainous areas. Over the years, they have already adapted well to the environmental conditions of their habitat. Generally, Chinese cymbidiums thrive in damp, sandy, and humus soil. They cannot withstand strong wind, heavy rain, snow, frost, heat, intense sun, high humidity, and arid conditions. For optimal growth of Chinese cymbidiums, it is necessary to have a good understanding of the biology of the plants as well as sound knowledge of the habitat and their growth conditions, which include humidity, temperature, sunlight, rainfall and soil conditions.

Sunlight is one of the most critical factors for plant growth. A plant depends on light energy to carry out photosynthesis. As discussed earlier, the light intensity requirement for photosynthesis varies with different kinds of orchids. Chinese cymbidiums are shade-loving plants, and they have comparatively low light requirements. However, if the sunlight is low, the plants will grow taller, and the leaves turn to a darker shade of green. The blooming rate and colour of the flower are greatly affected when grown under low light. The biosynthesis of flower pigments such as anthocyanins, like chlorophyll, is also dependent on light energy. Good ventilation is another important factor that cannot be ignored in the cultivation of Chinese cymbidiums. If the temperature and humidity in the orchid shed are high due to poor ventilation, most orchids will get infected by diseases easily. Hence, efforts should be made to prevent the shed from being "hot and stuffy".

Li *et al.*[60] have highlighted the following observations which are considered important for cultivating Chinese cymbidiums:

1. In the wild, Chinese cymbidiums normally grow well in leaf debris and products of their decomposition of primary/secondary forests in high mountainous areas. For example, *C. goeringii* grows on a mountain with rich soil derived from leaf debris with a pH of 5–5.6.

2. Chinese cymbidiums are shade-loving terrestrials. They prefer diffuse or morning light, relatively humid environment with well-drained soil, clean air and free air movement. Most Chinese cymbidiums require 50–70% shade in the summer. The light requirement varies with the different species:

   C. ensifolium, 60–70% shade,
   C. goeringii and *C. faberi*, 70–80% shade,
   C. sinense, about 85% shade.

3. Generally, Chinese cymbidiums are relatively cold tolerant but cannot withstand temperatures lower than 5–7°C during the winter season. Most Chinese cymbidiums grow well between 25°C and 28°C in summer. April to September is the best growing season for most species. Different Chinese cymbidiums have different temperature requirements for optimal growth. For example, the subtropical climate in the southern part of

China where many orchid nurseries are located is more suitable for *C. sinense* and *C. ensifolium,* but not *C. faberi* and *C. goeringii.* The latter two species grow well in regions where the climate is cooler.

4. It is recommended to keep the relative humidity at 70–80%. During the winter, Chinese *Cymbidium* enters partial dormancy and a relative humidity as low as 50% will not do much harm to them.

When one starts to grow Chinese cymbidiums, it is important to select a healthy plant with strong and fleshy roots. Diseased leaves and rotten roots must be removed. The plant must be washed thoroughly under a weak stream to avoid damage to the roots and leaves.

When planting Chinese cymbidiums, one should fill the bottom of the pot with a larger-sized potting medium (such as broken bricks) to about one-third of its volume and follow by the smaller ones. The plant that is being potted must not be placed too deep as this will hinder the emergence of new bud growth. Small pieces of potting materials should be placed on the top. Mild pressure should be applied at the top of the potting mix to ensure that the orchid is planted firmly. One should water the newly potted plants slowly until water starts dripping from the bottom of the pot. The plant should be placed in a shady location, and one should refrain from watering for a week to allow the plant to stabilise. It is a good practice not to move the orchid from one place to another too often once it is established, as the movement may hurt the roots.

It is a common practice to plant Chinese cymbidiums with 2–3 pseudobulbs together in a group, and at least one of them is a new shoot with 1–3 healthy roots. The timing of planting or transplanting depends on the species. Normally, it is best to start the planting of Chinese cymbidiums in autumn for those that will flower in spring, whereas for those that flower in autumn, they should be planted in spring.[60,64]

Flowerpot selection

Flowerpots for growing Chinese *Cymbidium* are generally deep (suitable for the growth of its long, and fleshy roots) and have a slightly enlarged trumpet-like mouth (easier to place the plants into the pot). The tall flowerpot often comes with three-corner support stands at the base. Some of these tall pots have many small holes drilled around them to promote ventilation and drainage (Fig. 5.1).

There are two main kinds of flowerpots: clay and porcelain. Clay flowerpots are cheaper but less presentable, whereas the porcelain flowerpots are more attractive and expensive, but allow less air exchange and permeability to water. Nowadays, people like to use plastic pots to grow orchids as it is not only cheap but also light and portable. It is important to select the right size and type of flowerpot for growing different Chinese *Cymbidium* species. For example, *C. sinense* which is relatively larger in size requires a bigger and taller pot. Medium-sized pots are preferable for *C. ensifolium* and *C. kanran*. Smaller pots are suitable for *C. goeringii*.[64]

It is interesting to note that selecting a well-designed flowerpot was an important consideration when growing Chinese cymbidiums for display in the past. This still holds true today. The exquisite

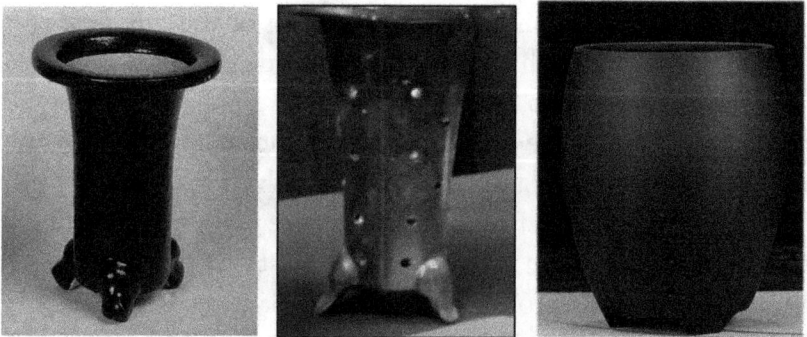

Fig. 5.1.   Tall flowerpots for growing Chinese cymbidiums. (Left and middle): Tall tripod pot with three supports at the corner of the stand base; (Middle and right): Tall pots with small holes to promote ventilation and drainage.

Fig. 5.2.   Flowerpots for Chinese *Cymbidium* with calligraphy and other designs.

design of the flowerpot is part and parcel of an overall appreciation of the beauty of the Chinese cymbidiums. In the past, wealthy orchid lovers and high government officials were very particular in the selection of orchid flowerpots and stands. They believed that only an elegantly designed flowerpot and stand would fit the exquisite character of Chinese cymbidiums. Therefore, they did not hesitate to spend money to engrave the customised flowerpots with Chinese calligraphy or exquisite paintings to reflect the status and demeanour of the wealthy families[47,54] (Fig. 5.2).

Potting media

One of the important things in the cultivation of Chinese cymbidiums is to choose a good potting mix for the plant, whether it is a cultivated variety or one collected from the wild. An ideal potting mix should have good water retention capacity and drainage that can provide a healthy environment for orchid root growth. The potting mix for growing Chinese cymbidiums is essentially the same as those for other terrestrial orchids. They consist of burnt soils, leaf mould, pond mud, tree fern fibre, sphagnum moss, tree bark, sand (coarse and fine), small broken brick or charcoal pieces.[60,64]

Burnt soil: This is a dried non-sticky potting medium that has been heat treated and sterilised. Due to its porous and granular structure, it provides good aeration for plant growth. Burnt soil is often mixed with compost, tree bark, or volcano sand to make a good potting mix for Chinese cymbidiums.

Leaf mould: It is the compost produced by the decomposition of fallen leaves in forests or gardens brought about primarily by a fungal breakdown. It is dark brown in colour, has a crumbly texture, and is a good additive to a potting mix. It is best to use leaf mould or soils collected from the wild where Chinese *Cymbidium* is growing. This could be related to the specific fungi associated with orchid roots that are present in the leaf mould or soils. Leaf mould promotes plant growth well initially, but because of its water retention ability, it could invite fungal or bacterial growth. Therefore, it is important to have the right proportion of leaf mould in a potting mix.

Pond mud: In Guangdong, many growers like to use year-old sun-dried pond mud as a potting substrate to grow *C. sinense* and *C. ensifolium*. Pond mud is good for the early growth of orchids. However, it soon disintegrates and forms crusts, which have an undesirable effect on root growth. In addition, because of its high nitrogen content, it will bring about significant changes not only in the leaf colour but also in the leaf form and structure of the plant, resulting in a loss in the value of its leaf art. Therefore, the use of pond mud as a potting medium in China has slowly been phased out.[60,64]

Tree fern fibre: It is the fibrous material harvested from the trunks of tree ferns. It has the advantage of not decaying easily, is porous, and provides good drainage and aeration. Its drawback is that it dries rather fast, and ants like to build nests in it.

Sphagnum moss: It holds water much better than other potting media. Its water retention ability makes it an excellent potting material for young orchids. However, it is necessary to ensure that it does not get too wet because that will reduce aeration, which in turn will inhibit root growth. Sphagnum moss is also used when displaying orchids in exhibitions. The moistness of the moss can be used as an indicator as to when to water the potted orchids.

Tree bark: It is common to include tree bark pieces in potting mix. It is also often used for decorative purposes and to help prevent weed growth. The tree bark is tough, and it does not decay easily. It has good water absorption and drainage properties and is conducive to root growth. However, once the bark decomposes, the process encourages fungal and disease growth which can harm the plants. In China, people use pine or longan tree bark for growing Chinese cymbidiums.[64]

Charcoal: Charcoal is another ideal potting material with known adsorption properties. It provides good drainage and ventilation, but lacks nutrients. It helps the potting mix to hold plant nutrients and can bind minerals such as phosphorous and calcium from the potting mix, making them more easily available to the plant. This reduces the leaching of nutrients following watering, and inhibits the growth of fungi and bacteria.

River sand, broken brick, and small stones: These potting materials allow good aeration and free water movement which are good for orchid roots growth. However, the water and fertilizer are lost quickly after application. On the other hand, they are stable and help to drain water exceptionally well and they are a good additive to any potting mix.

In different regions of China, people use different potting media to cultivate the Chinese cymbidiums. For example, in Jiangsu and Zhejiang areas, some growers like to use mountain soil to grow Chinese cymbidiums, whereas in Wuhan, it is customary to plant them with more than one-year-old cinder bark or burnt soil. Sugarcane is produced in the Pearl River Delta of Guangdong, and there is a large amount of bagasse available. After one or two years of retting and shredding, it has also been used as a Chinese *Cymbidium* potting medium. In Liancheng, Fujian, growers like to use a potting mixture consisting of tree bark, burnt soil, and peanut shells.[64]

Watering and fertilizer applications

In many aspects, the present cultivation practices of Chinese cymbidiums are not that different from that of the past. Like other orchids, the routine works involved in taking care of Chinese cymbidiums are watering, fertilizer application and pest control.

Developing a good watering program is important for the management of orchid cultivation. As with other plants, watering requirements in orchids depends on factors such as the weather, quality of water, the season, variety, stages of growth and location. For instance, one needs to control the relative humidity levels, maintain water quality and master the frequency and timing of watering.[60]

Chinese cymbidiums have fleshy roots that are known to withstand drought. They do not like to be over-wet as excessive watering encourages fungal and bacterial infections. Watering is done either by spraying or drenching. Generally, in Guangzhou, watering by spraying is 2 to 3 times per day except rainy days in spring and summer; while drenching is done once every 5–7 days. *C. sinense* needs more watering than *C. goeringii* and not surprisingly it is commonly found growing near a mountain stream in the wild.[60]

Types of fertilizer (organic and chemical) used for Chinese cymbidiums vary in different parts of China. According to Liu[64], many of the orchid growers use organic fertilizers, such as animal waste, barn waste and bone powder as they are easily available and of low cost. In mainland China, the soil and climate vary greatly from place to place so each region has its own methods to fertilize the plants. For example, in Zhejiang, some growers use a mixture of plant ash, bean cake, and animal bone ash to fertilize *C. goeringii*. In Fujian, *C. ensifolium* is planted with mixed burnt soil and dried stems and leaves of tree ferns. Guangdong farmers like to use pond mud, especially those that are black and oily. In Yunnan, Guizhou, and Sichuan, growers use humus-rich sludge as fertilizer to plant various types of Chinese cymbidiums.[64] Initially, growth is vigorous with the application of these organic fertilizers due to their high content of organic matter and slow-release properties. With time, organic fertilizers tend to break down and spoil the potting mix, leading to poor drainage and root rot. The practice of using plant/animal waste (e.g., bean cake, bone powder, cow dung) as organic fertilizers for growing Chinese cymbidiums is a low-cost option. However, the nutrient levels do vary considerably. In addition, the waste is often pungent smelling, and it also tends to promote vegetative plant growth.

Nowadays, many orchid gardens and nurseries in China are using chemical fertilizers which come with different NPK ratios, with or without trace elements.[65] Commercial fertilizer listed as 10–10–10 would mean equal content (10%) of nitrogen, phosphorus, and potassium is present in the makeup of the fertilizer. This is commonly referred to as a balanced fertilizer. Different stages of plant growth require fertilizers with different NPK ratios. A leaf tissue analysis of the NPK ratio of the plant could give us an idea of which element is low.

As discussed earlier, high nitrogen promotes leaf and root growth and is more suitable for the growth of a young plant. High potassium and phosphorus are recommended for flowering plants. Leaf tissues analysis of a flowering *C. sinense* has an NPK ratio of 6:1:9.[88] This indicates that there is a higher requirement for potassium for flower production. Extensive studies of fertilizer application practices, such as the effect of organic and inorganic fertilizer on tropical and epiphytic orchid growth, have been well documented.[14]

In addition, the controversy of adopting either foliar application or root feeding for fertilizer application has been reviewed.[14]

Penetration and uptake of the mineral through foliar application and root feeding have been confirmed with the use of radioactive phosphorus (^{32}P). The question that remains is the uptake efficiency between the two methods. To be able to fully exploit the potential of either foliar application or root feeding for optimal orchid growth, a good understanding of uptake, transport, storage, and reutilisation of minerals in orchid plants is needed. In China, foliar application of fertilizer is popular as it is more convenient and has a lower chance of harming the roots.[60]

Ideally, one should fertilize the orchids with lower amounts at more frequent intervals. This prevents excessive accumulation of nutrients in the potting media, wastage of fertilizer, and the growth of fungus and bacteria. Furthermore, Chinese cymbidiums like other orchids have a slow rate of mineral (nutrient) uptake.[27] It is a common practice in China to apply fertilizer to orchids in the morning in spring and autumn. In summer, it is best done in the late afternoon as this prevents the formation of nutrient droplets on the leaves, which may act as small lenses that focus the sunrays on the leaves and thereby cause leaf burn. It is also not advisable to apply fertilizers to the emerging flower buds, newly potted plants and plants undergoing dormancy.[60]

Pests and diseases

Pests

Chinese cymbidiums like many other crops and ornamental plants are attacked by various kinds of pests and diseases. The common pests in orchid cultivation found in southern China can be broadly divided into two groups[60,65]:

- Insect pests: aphids, scale insects, mealybugs, thrips, and ants.
- Non-insect pests: red spider mites, snails, and slugs.

Aphids: These are small sap-sucking insects with fleshy, and soft bodies. They come in many colors, such as white, dark brown, green or red. They attack new growths and flower buds causing severe damage to the plant parts which become disfigured. Aphids are known to transmit virus diseases from plant to plant. The honeydew excreted by aphids attracts ants and is an ideal medium for the growth of sooty mould. Female aphids can breed live aphids asexually and at an extremely fast rate. Within a week, a single female can produce hundreds of offspring. This could cause great damage to the entire greenhouses or gardens. In Guangzhou, aphids are controlled by spraying with 20% Dimethoate (diluted 500 times).[60,65]

Scale insects: The soft scale, armoured scale, and mealybug belong to a closely related family. Soft scale is small, 0.31–1.25 cm in diameter, flat, oval, and pale yellow or brown. The shell of the soft scale is characterised by a thick waxy coating. It extracts plant fluids from the phloem by sucking fluids through its straw-like mouthparts. This will reduce plant growth and cause the leaf to drop. Armoured scales are also small, about 0.16–0.31cm in length, flat and circular shaped. They are found on leaves, twigs and branches. They can also cause injury by sucking sap from young leaves and buds. Spraying with 40% dimethoate (diluted 1000 times) provides effective control against scale insects.[65]

Mealybugs: These are small, oval, sap-sucking insects, 0.25–0.63 cm in size, that secrete white powdery wax substance as a protective coating, making it easy to identify when they appear on the stems and leaves (leaf axils) of plants (Fig. 5.3). Mealybugs excrete substances like honeydew which encourages the growth of sooty mould. Their feeding damage can be quite severe, including wilting and distortion of new leaves but mealybugs can be gently scraped off.

Thrips: Thrips are minute sucking insects, with slender bodies, fringed wings, and possess distinctive asymmetrical mouthparts, which feed on young flowers. Infested flowers can often become

Fig. 5.3. Mealybugs form a powdery and waxy layer on stems and leaves of Chinese *Cymbidium.*

Fig. 5.4. Thrips damage the flower buds of Chinese *Cymbidium,* causing distortion and discoloration.

deformed (Fig. 5.4). Managing thrips is more difficult as they always hide in flower buds, making it hard to spot them and for pesticides to reach them. The young thrips are not only minute but also colourless. The use of a systemic pesticide has been suggested as one mean of controlling thrips.

Mites: They are tiny, almost invisible, and are members of the Arachnid (spider) family. Red spider mites always hide on the lower

Fig. 5.5.   Red spider mites damage leaf of Chinese *Cymbidium*, causing chlorotic spots on the lower surface of the leaves.

side of the orchid leaves, making it difficult to detect and eradicate them (Fig. 5.5). They feed by using the sharp mouthpart to pierce through the leaf epidermis into the mesophyll and palisade tissues, sucking up valuable plant nutrients and chloroplasts thus causing chlorotic spots on the lower surface of the leaves. One can try wiping or spraying with dilute detergent on the upper and lower sides of the leaves, but this is slow and tedious. Under warm temperatures, mites multiply rapidly. If the infection is serious, spraying with miticide is recommended.[60]

Snails and slugs: They like to hide in damp and dark places during the day and come out at night feeding on the root tips, buds, young leaves and flowers. In the olden days in China, the method of controlling the snails and slugs was to lure them out using apple peels as bait or to use beer to make them intoxicated, thus resulting in death. Now they can be easily eradicated with slug pellets.

Fungal and bacterial diseases

The importance of controlling the pests and diseases in Chinese cymbidiums should not be taken lightly. Unlike many other flowering plants, one appreciates the beauty of a Chinese *Cymbidium* for its flowers (flower art) and the leaves (leaf art) as well. Disease and pest infestation will not only affect the growth of the plants but also seriously reduce the ornamental value of some Chinese cymbidiums where the focus is on leaf art appreciation.

Chinese cymbidiums, like other plants, are attacked by bacteria, fungi and viruses. The incidence of diseases depends heavily on environmental conditions and the management of gardens and nurseries. Orchid plant diseases often occur when excess moisture stays on the leaves and flowers, or on the growing medium with poor drainage. The common orchid diseases can be prevented if plant health is monitored frequently, and diseases are acted upon as soon as they appear.

The common diseases of Chinese cymbidiums are of fungal or bacterial origin. Generally, the diseases occur in roots, pseudobulbs, and leaves and usually cause some form of rot, discolouration, and spotting in flowers and leaves. Some of the common fungal and bacterial diseases of Chinese cymbidiums and their control in southern China are described as follows[60]:

Anthracnose: Anthracnose is one of the most common fungal diseases of orchids and it infects the developing shoots and leaves. The first symptom of an anthracnose infection is the presence of yellowish-brown lesions on the leaf. These lesions enlarge with time and turn dark brown and cause the infected leaves to rot and die (Fig. 5.6). In

Fig. 5.6.　Leaf damage in Chinese *Cymbidium* due to anthracnose, a fungal disease.[60]

southern China, infections usually occur mainly in July and August when humidity is high, and the weather is hot. It is important to keep the place well ventilated to reduce infection. Increasing the application of phosphorus and potassium fertilizers would enhance the disease resistance of the plants. Spraying 2–3 times at 10–15-day intervals with a fungicide such as 50% Carbendazim (diluted 800 times) provides good control against the infection.[60,65]

Southern blight: It is another fungal disease that is found quite commonly in *C. sinense* and *C. ensifolium* in southern China. The fungal pathogen infection is most prevalent when the weather is warm and humidity is high, especially during the rainy reason. Mycelia first attack the basal part of pseudobulbs, roots, leaves, and the lower part of the stems. The infested surfaces are covered with a layer of silk-like mycelia which soon turn brown or black (Fig. 5.7). The disease soon spreads to the entire plant, and this is followed by rapid plant death.

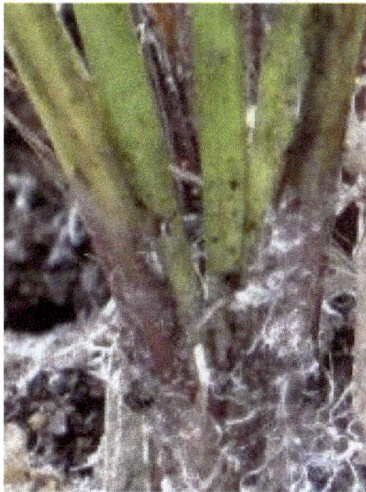

Fig. 5.7. Southern blight is a fungal disease which attacks the basal part of pseudobulbs, roots and leaves of Chinese *Cymbidium*.

The occurrence of the southern blight disease can be prevented using sterilized planting materials and avoidance of heavy use of fertilizers. The infected plant must be washed with running water and dipped for 20 minutes in fungicide such as 50% Carbendazim (diluted 1,000 times). Fungicides should also be applied to the soil in and around the infected areas as this will inhibit the growth of the fungus.[60,65]

Bacterial soft rot: Bacterial soft rot that damages the orchid plants is caused by a bacterial infection that can occur under hot and moist conditions. Infected leaves develop water-soaked areas often with brown halos. The pathogen spreads quickly, first to the roots and pseudobulb, then to the rest of the plant. Rotten areas emit an unpleasant foul smell. To control the spread of the disease, one should isolate the infected plants, avoid overwatering, and provide good aeration for the plants. Additionally, spraying the plant with 1% Bordeaux has been reported to control the disease as well.[60]

Root rot: The disease often occurs when the potting mix is waterlogged due to over-watering, poor drainage, or over-fertilising. The fungus (*Fusarium sp.*) would first invade the roots, causing the roots to rot gradually and eventually causing the whole plant to die. If the infection is detected early, change the pot quickly, pour out all the potting media, and cut off the rotten roots and leaves. To get rid of the disease, one should soak the whole plant in a 0.1% potassium permanganate solution for about 15 minutes, wash and let it air-dry, and replant it in a new potting mix. The plant should be placed in a ventilated area and be allowed to gradually regain its vitality.[60]

Viral diseases

A high percentage of orchids are afflicted with viral diseases. Viral diseases are systemic in nature, and they can invade all cellular parts of the plant. *Cymbidium mosaic virus* (CymMV) *and Odontoglossum ringspot virus* (ORSV) are two of the economically important viruses affecting orchids worldwide.[21] According to Chen & Ji[65] in China, viral infection is common in cultivated orchids, but it is rarely seen in wild orchids.

Viruses are spread by insects and humans. The human is now believed to be the main facilitator for the spread of viral diseases. It is transmitted through our cutting and repotting activities. It is good that all used pots or utensils (knife, scissor, cutting blade) be thoroughly sanitised to prevent virus spread. It is a general belief that an orchid infected with a virus is not curable. Once infected, the orchids must be destroyed or burned immediately. However, there is a way to remove the virus from orchids, but it will be costly and time consuming. This involves the selection of appropriate tissues culture techniques and chemical treatment for the elimination of virus in orchids. A high percentage of virus-free plantlets, for example, could be obtained from the thin-section tissue culture of infected orchids treated with ribavirin.[24]

Viruses usually cause yellow or brown mottling and streaking of the leaves, or colour breaks in the flowers and there is a reduction in growth (Fig. 5.8). The symptoms vary from plant to plant, and from location to location and even purchasing certified virus-free plants will not ensure that they are totally virus free. In recent years, encouraging progress has been made in the application of the

Fig. 5.8.   Chinese *Cymbidium* leaf infected by virus.[60]

pathogen-derived resistance concept to generating orchids with resistance to viral infection.[21]

In short, pests and diseases in orchids except viral diseases can be controlled using appropriate insecticides, bactericides or fungicides. Pesticide can either be contact or systemic. In contrast to contact pesticides, the systemic pesticide is designed for the chemicals to be taken up and transported to all parts of the plants, thereby protecting the whole plant from diseases and pests. Extreme care should be exercised in using any of these agents on human and pet dwellings.

In recent years, it has been a general trend to reduce the use of chemical pesticides worldwide both for the safety of the user and the protection of the environment. In China, some Chinese *Cymbidium* growers prefer to use natural pest control methods instead of using chemical pesticides. For example, they cut the stems of the *Artemisia argyi* (commonly known as Chinese mugwort) (艾) into 3-cm-long pieces and put them on the top of the potting media to control aphids and spider mites. Other growers use peeled purple garlic. The garlic is mashed, mixed with water and sprayed onto the leaves and soil to control scale insects and aphids. A garlic clove buried in the soil can also drive away ants and nematodes. Some hobbyists like to make their own non-toxic, natural organic pesticides by soaking tea cake, cinnamon powder, and cigarette shreds in water to control pests and diseases.

We have provided a general account of pests and diseases and their control in the cultivation of the Chinese cymbidiums. Readers are encouraged to consult more advanced and specialised handbooks to gain more in-depth information on this subject.[6,10,65]

Chapter 6

Prospects for Chinese *Cymbidium*

In the last forty years, phenomenal economic development and progress have been made in China. The Gross Domestic Product (GDP) in China is now ranked as one of the highest in the world and the standard of living of middle-class families has improved tremendously, thus changing the lifestyle of the Chinese. People can now lead and enjoy a higher quality of life and indulge in their love of *lan*. As a result, it is now customary for *lan* lovers to invite a few close friends to a serene place to have tea while appreciating the beauty and elegance of their Chinese cymbidiums and sharing their cultivation experience. The love and the appreciation for Chinese cymbidiums have seeped into every aspect of social life in China and have become an integral part of Chinese culture.

6.1. Commercial Cultivation of Chinese *Cymbidium*

According to historical records, Chinese *Cymbidium* shows started during the Qianlong period (乾隆; 1736–1796).[47] The flower shows first took place at the City God Temple of Shanghai (上海城隍庙) around March and April every year. Usually, about 30 orchid lovers and growers would take part in the show. Soon thereafter, the flower shows were also staged in Suzhou and Wuxi. Orchid lovers from Ningbo, Shaoxing, Nanjing, Jiaxing and Yixing travelled hundreds of miles by boat just to showcase and sell their plants at the shows held in Shanghai, Suzhou or Wuxi. Over time, venues for flower shows started to expand from the City God Temple of Shanghai to county guild halls, private gardens, teahouses, and restaurants.[47] As Chinese orchid flower shows became much better organised, the exchange and trading of orchids became prevalent amongst orchid growers and lovers.

Similar to other countries, Chinese cymbidium sales in China today are mainly through two channels: the hobbyists' markets and the consumers' markets. The hobbyists' market is small because it caters especially to those who seek exquisite and unique Chinese cymbidiums. However, while the hobbyists' market has shrunk, the general consumers' market has expanded quickly.

The sale of Chinese cymbidiums depends on supply and demand and has been heavily influenced by the media. The trading of Chinese cymbidiums is now usually conducted via garden shows, exhibitions (Fig. 6.1), local flower markets, or via online platforms. A large flower exhibition centre with a special hall for the display of Chinese cymbidiums and Western orchids was built in Wengyuan (翁源), Guangdong, in 2018 for RMB 17 million, covering an area of 50,000 sq. m., further facilitating the sale of the Chinese cymbidiums.[46]

Fig. 6.1.   Chinese *Cymbidium* exhibition.

Many orchid societies in China form online groups on social networking platforms, where they can trade and share their appreciation of orchids, including Chinese cymbidiums. China's orchid network is widely distributed in various provinces such as Guangzhou, Zhejiang and Yunnan. For instance, the "China Orchid Trading Network" (中国兰花交易网)[52] was founded in 2003. This network offers a reliable, practical and efficient online orchid platform, providing detailed information regarding orchid plants and flower sales. This includes orchid auctions with a comprehensive and timely display of orchid varieties, prices, trading statistics, feedback of sale transactions, orchid friends' communities, orchid garden shops and associations. As of the first quarter of 2011, the "China Orchid Trading Network" listed more than 150,000 members which basically include most of the online orchid shopping population in China. This network has become a popular online platform for orchid amateurs, growers, and traders to interact and learn more about market trends, recent advances in orchid cultivation, the latest scientific and technological achievements, and the development of the orchid industry.

After the first China Orchid Expo held in Guangzhou in 1988, greater public awareness of the commercial potential of Chinese orchids began to take shape. The commercial cultivation of Chinese cymbidiums has developed rapidly in recent years, especially in the southern provinces with rich resources of Chinese cymbidiums. These provinces include Jiangsu, Zhejiang, Guangdong, Fujian, Yunnan, Guizhou and Sichuan. In the early 1990s, Guangdong and Fujian began large-scale production of Chinese cymbidiums and started to export them to Japan, South Korea, Taiwan, Hong Kong and Macao. Presently, China has four major Chinese orchid regional production markets — Zhejiang, Sichuan, Yunnan and Guangdong.[44,63]

Today, many high-class hotels, conference halls and government offices in China display potted Chinese cymbidiums to project a distinguished and elegant image. This could be influenced by the popularity of displaying *Phalaenopsis* as a decorative potted orchid in China, the USA, and other places all around the world. *Phalaenopsis* flowers are showy, have a long life, and when grown in a group of 3–5 plants, make a very impressive sight. While Chinese cymbidiums are not as attractive as *Phalaenopsis*, they certainly are unique and have a charm of their own.

In recent years, big enterprises are investing not only in the *Phalaenopsis* but also in Chinese cymbidiums. Due to the high land cost and less favourable environmental conditions in and around the Pearl River Delta (珠江三角洲) region, many of the orchid farms in the Chen Chun area (陈村) are moving to Yunnan, Guizhou and other locations where land is cheaper and the environmental conditions are more favourable for orchid production.[45]

Liu[63] highlighted that, compared to other countries, there is still considerable room for further improvement and expansion in the orchid industry in China, including Chinese cymbidiums. The major constraint is the level of commercialisation, management, and marketing of orchid production in China that makes it challenging to meet the growth and requirements from market demand. Another issue facing the future development in the Chinese orchids industry is the urgent need to breed more new varieties of excellent and unique orchids. As pointed out by Liu, while China has many excellent orchid germplasm resources, germplasm protection research and species selection have yet to be further improved.

Orchid cultivation has generally benefited greatly under the China High-Tech Development Plan. With research support from the government under this scheme, significant progress has been made in the hybridisation and breeding programme, mass propagation, molecular biology, cultivation technology and pest control of Chinese *Cymbidium* orchids. Taking advantage of the improvements made in production technology and management reform, commercial companies with the vision and strength have taken the lead in the development of large-scale commercial orchid cultivation. With the investment of big enterprises and modern administration, the traditional management of the Chinese orchid industry has improved substantially. In 2015, the market size of the Chinese orchids industry was RMB 22.24 billion, but it has increased to RMB 30.73 billion in 2020.[63]

6.2. Conservation of Chinese *Cymbidium* Species

Following the economic reform that opened up China to the world in the 1980s, there was a renewed interest in Chinese *Cymbidium* culture, creating a great demand for new varieties of Chinese *Cymbidium*. There was also a rush to collect unusual wild Chinese cymbidiums, creating great price speculation for these rare orchids.[30] These overharvest trends combined with rapid development of the orchid industry have caused great concern amongst nature lovers and environmentalists, raising awareness on the need to conserve Chinese cymbidiums.

Considerable effort has been made to conserve the wild Chinese cymbidiums and other orchids.[16,25,29,30] The conservation effort of orchid diversity in China over the last 21 years (2000–2020) has been reviewed recently.[43] This study has provided detailed and updated information on orchid biodiversity and useful suggestions for orchid conservation in China. It showed that of the 1502 known native orchid species, 653 species were identified as facing the risk of extinction and four species endemic in China were already classified as extinct. Additionally, it was estimated that approximately 90% of the 653 threatened orchid species had been affected by habitat loss and deterioration. Monitoring population dynamics and *in-situ* conservation are the two major tasks recommended for conserving the endangered Chinese orchid species in the near future.[43]

To protect orchid resources in China, the Chinese government has established a large number of nature reserves and botanical gardens as bases for introducing and cultivating rare and nearly extinct orchid species.[40] According to Zhou *et al.* (2021), in the last six decades, 474 national reserves for *in-situ* conservation of plant diversity including orchids have been established in China. About 1111 native orchid species are protected in national nature reserves, with another 66 species protected in provincial nature reserves. The government has also designated certain rare orchid species as endangered plants under state protection. Advancement in mass clonal propagation of quality Chinese orchids and *in-situ* studies of wild orchid mycorrhizal relationship also help to assist the successful restoration and sustainable use of orchids including Chinese cymbidiums in China.[26,35] In recognition of the rapid decline of wild orchid populations in China, and to support the global strategy for plant conservation, the Chinese government introduced a strict export ban on wild orchids in 2008.[30]

A dramatic shift in the *Cymbidium* markets in China was observed in the 2010s. Nearly all the shops in Sichuan were moving away from selling wild orchids and started selling greenhouse-produced orchids of many genera. One contributing factor was the significant advances made in the micropropagation of orchids. The shift in purchasing preference for tissue-cultured Chinese cymbidiums by the younger generations is noteworthy. Quality clonal Chinese cymbidiums can now be easily purchased at relatively affordable prices. This has also reduced the collection pressure of wild Chinese orchids.[30] These developments bode well for the future of Chinese cymbidiums both for hobbyists and consumers, but more importantly support the drive for its conservation.

China's orchid industry continues to advance with improvements in cultivation technology and rapid market growth. This holds endless growth potential and provides the opportunity for further market expansion. Coupled with the long history of Chinese *Cymbidium* culture, and with the help of the Internet, more and more people are starting to know and develop a love of Chinese *Cymbidium*. With all this, it is clear the development prospects for Chinese *Cymbidium* are bright and promising, and the "*Lan Culture*" will continue to flourish for time to come.

References

1. Arditti, J. (1977). Clonal propagation of orchids by means of tissue culture — manual. In: Arditti, J. (ed.), *Orchid Biology: Reviews and Perspectives I*. Cornell University Press, Ithaca. pp. 203–394.
2. Arditti, J. (1992). *Fundamentals of Orchid Biology*. John Wiley & Sons, New York.
3. Chase, M., Christenhusz, M. & Mirenda, T. (2017). *The Book of Orchids: A Life-Size Guide to Six Hundred Species from Around the World*. Ivy Press, Brighton.
4. Chen, S. C. & Tang, T. (1982). A general review of the orchid flora of China. In: Arditti, J. (ed.), *Orchid Biology: Reviews and Perspectives II*. Cornell University Press, Ithaca. pp. 39–81.
5. Chen, Y. Q., Liu, X. & Liu, Y. Q. (2005). In vitro plant regeneration from the immature seeds of *Cymbidium faberi*. *Plant Cell Tissue and Organ Culture* 81: 247–251.
6. Chuo, S. K., Ernst, R., Arditti, J. & Hew, C. S. (1994). Orchid pests — a compendium. In: Arditti, J. (ed.), *Orchid Biology: Reviews and Perspectives VI*. John Wiley & Sons, New York. pp. 455–563.
7. DuPuy, D. & Cribb, P. (1988). *The Genus Cymbidium*. Timber Press, Portland.
8. Eight Eccentric of Yangzhou. https://en.m.wikipedia.org/wiki/Eight_Eccentrics_of_Yangzhou. Retrieved 26 May 2022.
9. Gong, M. F., Guan, Q. L., Lin, T. X., Lan, J. M. & Liu, S. (2018). Effects of fungal elicitors on seed germination and tissue culture of *Cymbidium goeringii. AIP Conference Proceedings*, Vol. 1956, issues 1.

10. Hadley, G., Arditti, M. & Arditti, J. (1987). Orchid diseases — A compendium. In: Arditti, J. (ed.), *Orchid Biology: Reviews and Perspectives, IV*. Cornell University Press, Ithaca. pp. 261–340.

11. Hew, C. S. (1987). Respiration in orchids. In: Arditti, J. (ed.), *Orchid Biology: Reviews and Perspectives, IV*. Cornell University Press, Ithaca. pp. 227–259.

12. Hew, C. S. (2001). Ancient Chinese orchid cultivation: A fresh look at an age-old practice. *Scientia Horticulturae* 87: 1–10.

13. Hew, C. S. & Ng, C. K. Y. (1996). Changes in mineral and carbohydrate content in pseudobulbs of the C_3 epiphytic orchid hybrid *Oncidium* Goldiana at different growth stages. *Lindleyana* 11: 125–134.

14. Hew, C. S. & Yong, W. H. (2004). *The Physiology of Tropical Orchids in Relation to the Industry*. 2nd ed. World Scientific, Singapore.

15. Hew, C. S. & Yong, W. H. (2006). Orchids in Chinese medicine. *Innovation* 6: 2–4.

16. Hinsley, A., de Boer, H., Fay, M. F., Gale, S. W., Gardiner, L. M., Gunasekara, R. S., Kumar, P., Masters, S., Metusala, D., Roberts, D. L. (2017). A review of the trade in orchids and its implications for conservation. *Botanical Journal of the Linnean Society* 186: 435–455.

17. Holttum, R. E. (1964). *A Revised Flora of Malaya Vol.1: Orchids of Malaya*. 3rd ed. Government Printing Office, Singapore.

18. Hong, H. & Soh, S. B. (2020). *Healing Orchids*. World Scientific, Singapore.

19. Hu, S. Y. (1971). The Orchidaceae of China, I. Orchids in the life and culture of the Chinese people. In: *Quarterly Journal of the Taiwan Museum*, Vol. 24, pp. 72–101.

20. Huang, Y., Li, F., Xia,Y. & Chen, K. (2011). Scent profiling of *Cymbidium ensifolium* by electronic nose. *Scientia Horticulturae* 128: 306–310.

21. Koh, K. W. & Chan, M. T. (2017). Virus resistance in orchids. In: Chen, W. H. & Chen, H. H. (eds.), *Orchid Biotechnology, III*. World Scientific, Singapore. pp. 189–221.

22. Law, S. M. (Translated by T. Blishen) (2016). *Reading Chinese Painting: Beyond Forms and Colors, a Comparative Approach to Art Appreciation*. Shanghai Press.

23. Leung, P. C. (2020). *Gastrodia elata.* In: *Healing Orchids.* World Scientific, Singapore. pp. 27–59.

24. Lim, S. T., Wong, S. M. & Goh C. J. (1993). Elimination of cymbidium mosaic virus and odontoglossum ringspot virus from orchids by meristem culture and thin section culture with chemotherapy. *Annals of Applied Biology* 122: 289–297.

25. Liu, H. P., Liu, Z. J., Jin, X. H., Gao, J. Y., Chen, Y., Liu, Q. & Zhang, D. Y. (2020). Assessing conservation efforts against threats to wild orchids in China. *Biological Conservation* 243.

26. Liu, H. X., Luo, Y. B. & Liu, H. (2010). Studies of mycorrhizal fungi of Chinese orchids and their role in orchid conservation in China — A review. *The Botanical Review* 76: 241–262.

27. Pan, R. C., Ye, Q. S. & Hew, C. S. (1997). Physiology of *Cymbidium sinense*: A review. *Scientia Horticulturae* 70: 123–129.

28. Poole, H. A. & Sheehan, T. J. (1982). Mineral nutrition of orchids. In: Arditti, J. (ed.), *Orchid Biology: Reviews and Perspectives, II.* Cornell University Press, Ithaca. pp. 195–212.

29. Seyler, B. C., Gaoue, O. G., Tang, Y. & Duffy, D. C. (2019). Understanding knowledge threatened by declining wild orchid populations in an urbanizing China (Sichuan). *Environmental Conservation* 46: 318–325.

30. Seyler, B. C. & Tang, Y. (2020). Orchids, culture & markets: A brief history of China's wild *Cymbidium* trade. Orchid Specialist Group, Global Trade Programme.

31. Tan, H. T. W. & Hew, C. S. (1995). Rev. ed. *A Guide to the Orchids of Singapore.* Singapore Science Centre, Singapore.

32. Tanaka, M. (1991). Disposable film culture vessels. In: Bajaj Y. P. S. (ed.), *Biotechnology in Agriculture and Forestry, Vol.17, high-tech and Micropropagation I.* Springer-Verlag. pp. 212–228.

33. Teoh, E. S. (2016). *Medicinal Orchids of Asia.* Springer, Switzerland.

34. Tsuji, K. & Kato, M. (2010). Odor-guided bee pollinators of two endangered winter/early spring blooming orchids, *Cymbidium kanran* and *Cymbidium goeringii*, in Japan. *Pant Species Biology* 25:249–253.

35. Wu, J. R., Ma, H. C., Lu, M., Han, S. F., Zhu, Y. Y., Jin, H., Liang, J. F., Liu, L. & Xu, J. P. (2010). Rhizoctonia fungi enhance the growth of the endangered orchid *Cymbidium goeringii*. *Botany* 88: 20–29.

36. Wu, J. R., Ma, H. C., Xu, X. L., Qiao, N., Guo, S. T., Liu, F., Zhang, D. H. & Zhou, L. P. (2013). Mycorrhizas alter nitrogen acquisition by the terrestrial orchid *Cymbidium goeringii*. *Annals of Botany* 111: 1181–1187.

37. Xu, Q., Wang, S. T., Hong, H. Z. & Zhou, Y. (2019). Transcriptomic profiling of the flower scent biosynthesis pathway of *Cymbidium faberi* Rolfe and functional characterization of its jasmonic acid carboxyl methyltransferase gene. *BMC Genomics* 20: 125.

38. Xu, X. W., Chai, M. L., Yang, Y. P., Pan, K. K., Zeng, A. P., Jiang, N. & Wang, L. F. (2011). Hybridization and in vitro seed germination of *Cymbidium kanran*. *Acta Horticulturae Sinica* 38: 2010–2016.

39. Yam, T. W. & Arditti, J. (2017). *Micropropagation of Orchids*. 3rd ed. John Wiley & Sons, New York.

40. Yang Z. H., Zhang, Q. T., Feng, Z. Z., Lang, K. Y., & Li, H. (1993). *Orchids*. (Translated into English by Xiong, Z.R.) Kunming Institute of Botany, Chinese Academy of Sciences and Yunnan Academy of Forestry, Yunnan.

41. Yu, X. H., Luo, Y. B., & Dong, M. (2008). Pollination biology of *Cymbidium goeringii* (Orchidaceae) in China. *Journal of Systematics and Evolution* 46: 163–174.

42. Zhang, N. G., Yong, J. W. H., Hew, C. S. & Zhou X. (1995). The production of cytokinins, abscissic acid and auxin by CAM orchid aerial roots. *Journal of Plant Physiology* 147: 371–377.

43. Zhou, Z. H., Shi, R. H., Zhang, Y., Xing, X. K. & Jin, X. H. (2021). Orchid conservation in China from 2000 to 2020: Achievements and perspectives. *Plant Diversity* 43: 343–349.

44. 广东国兰产业现状和发展前景 (2017). (顺德农业信息网).

45. 广东顺德陈村国兰现状介绍 (2015). http://www.flowerworld.cn/. Retrieved 26 May 2022.

46. 广东翁源建成中国兰花博览中心投资约1.7亿元 (2018).

47. 马性远, 马扬尘 (2008) 中国兰文化. 北京: 中国林业出版社.

48. 马鑫 (2021). 寒兰: 冷香扑鼻, 风姿绰约. https://baike.baidu.com/tashuo/browse/content?id=77a06242008626d606bea91b. Retrieved 26 May 2022.

49. 王汇涛 (2008). 范画临本 — 画兰. 南京: 江苏美术出版社.

50. 王传贺 (2014). 怎样画兰. 北京: 中国书店.
51. 乌纱帽 https://baike.baidu.com/item/乌纱/6680229. Retrieved 26 May 2022.
52. 中国兰花交易网 https://baike.baidu.com/item/中国兰花交易网/8859019. Retrieved 26 May 2022.
53. 中国兰花邮票Baike.baidu.com/item/中国兰花/23216080. Retrieved 26 May 2022.
54. 中国兰蕙. 台北: 雷鼓出版社.
55. 中国栽兰史--历代兰谱经典著作 (2019). https://m.sohu.com/a/301743544_278321/?pvid=000115_3w_a. Retrieved 26 May 2022.
56. 王贵学 (1247; 宋) 王氏兰谱 (Wang Shi Lan Pu) 香艳丛书(卷一).
57. 王家兴 (2016). 兰魂. 北京: 中国文史出版社.
58. 卢思聪 (1990). 兰花栽培入门. 北京: 金盾出版社.
59. 兰谱奥法. 说郛 (卷六十三) (涵芬楼排印本).
60. 李少球, 胡松华, 鲁章 (1995). 中国兰花 (品种、欣赏、栽培). 广州: 广东科技出版社.
61. 许东生 (2011). 国兰名品赏鉴 (上, 下册). 北京: 中国林业出版社.
62. 扬州八怪 https://baike.baidu.com/item/扬州八怪/335. Retrieved 26 May 2022.
63. 刘伟 (2021). 中国兰花行业市场深度调研及投资前景研究预测报告.
64. 刘清涌 (1991). 兰花. 广州 : 广东旅游出版社.
65. 陈心启, 吉占和 (1997). 中国兰花全书. 北京: 中国林业出版社.
66. 佛兰 http://yao51.com/jiankangtuku/pwggmhmky.html. Retrieved 26 May 2022.
67. 求兰花与佛教的联系, 典故 (2012). https://zhidao.baidu.com/question/419426455.html. Retrieved 26 May 2022.
68. 吴应祥 (1991). 中国兰花. 北京: 中国林业出版社.
69. 吴应祥 (1990). 中国兰花漫谈 (自: 中国兰花. 何清正, 陈心启编著) (2–3 页) 成都: 四川美术出版社.
70. 陆越子 (2000). 画兰要诀. 南京: 江苏美术出版社.
71. 国兰矮种 https://www.hua.com/baike/FamousFlower/mh026.htm. Retrieved 26 May 2022.
72. 郑板桥纪念馆 https://baike.baidu.com/item/郑板桥纪念馆/2504812. Retrieved 26 May 2022.

73. 郑板桥邮票 http://www.518yp.com/youpiaotupian/6806.html. Retrieved 26 May 2022.

74. 郑板桥画选 (1989). 北京: 荣宝斋编辑出版社.

75. 周积寅 (1984). 郑板桥书画 (前言). 山东省文物局, 潍坊地区出版办公室编著. 山东美术出版社 (1–4页).

76. 郑鸿 (2000). 老子思想新译. 八方文化企业公司.

77. 胡老溪 (2013). 中国画名家。江苏美术出版社.

78. 赵时庚 (1233: 宋) 金漳兰谱 (Jin Zhang Lan Pu). 香艳丛书 (卷一).

79. 郭林生 (2019). 远离烦恼心. 新北市: 木马文化出版社.

80. 荆棘丛兰图https://baike.baidu.com/item/荆棘丛兰图/24279696. Retrieved 26 May 2022.

81. 难得糊涂   www.jubaozhai.com/arts_1.asp?id=206&typename=. Retrieved 26 May 2022.

82. 清代鼻烟壶. http://collection.sina.com.cn/wwzx/20131219/16211 37605.shtml. Retrieved 26 May 2022.

83. 崔庆忠 (2003) 图说中国绘画史. 台北市: 扬智文化事业股份有限公司.

84. 培育奇花 根子就在根儿上 (25 March 2014). https://www.youtube.com/watch?v=kTNpFtSTBCQ. Retrieved 26 May 2022.

85. 鹿亭翁 (宋) 兰易. 四明丛书 (海宁蒋氏别下斋钞校本).

86. 道教文化与兰花 (2021). https://www.mlbaikew.com/html/51735.htmlb. Retrieved 26 May 2022.

87.《魅力兰花》之兰花赏析 (2020). https://m.sohu.com/a/399410976_678607/?pvid=000115_3w_a. Retrieved 26 May 2022.

88. 潘瑞炽, 叶庆生 (2006). 国兰生理. 北京: 科学出版社.

89. 儒学的核心是什么 (2016). http://m.edu.iask.sina.com.cn/bdjx/6iXL 1Z6ZXxE.html. Retrieved 26 May 2022.

Index

The Origin and Meaning of
"*Lan*", 3
tian ma, 7
translocation, 114

umbilical cord, 116

velamen, 90
viral diseases, 140
virus-free plants, 141

Wang Guixue, 11
Wang Shi Lan Pu, 11, 87, 121
water relation, 117
watering, 123, 132
Wen Zhengming, 30
Wu Changshuo, 33
Wu Enyuan, 13

xiao ren, 23
Xu Wei, 25, 30

Yang Kui, 8
Yong Lan, 8
You Fang Ting, 8
Yulan, 24

ze lan, 6
Zen Flower, 19
Zhang Heng, 22
Zhao Mengjian, 25, 26
Zhao Shigeng, 10, 12
Zheng Banqiao, 25, 35, 36, 39, 44, 46
Zheng Sixiao, 27
zhi lan, 4, 6
Zhi Lan Shuo, 8
Zhou Lv Qing, 12
Zhu Kerou, 13

www.ingramcontent.com/pod-product-compliance
Lightning Source LLC
Chambersburg PA
CBHW050629190326
41458CB00008B/2197